To Ir. ' Dr. Zubair,

Eat well + smile
often!

DENTAL 101

3/4/ 11

Eat well + smile

DENTAL 101

THE 101 FACTORS YOU SHOULD KNOW BEFORE GOING TO THE DENTIST

UMAR HAQUE, DMD

OBS PRESS
OAKBROOK TERRACE, ILLINOIS

OBS Press
A Division of Oak Brook Smiles, PC
1S132 Summit Avenue, Suite 200
Oakbrook Terrace, Illinois 60181

For information regarding special discounts for bulk purchases,
please contact OBS Press at 1-630-627-7420 or
info@oakbrooksmiles.com

Paperback ISBN: 978-0-557-45614-7
Hardcover ISBN: 978-0-557-45613-0

In The Name of God, The Most
Merciful, The Most Kind

To my children, Maaz and Sana:
May God help you achieve your full
potential.

TABLE OF CONTENTS

INTRODUCTION 1

CHAPTER 1: FACTORS 1–10
TOP 10 REASONS TO GO TO THE DENTIST 5

CHAPTER 2: FACTORS 11–20
TOP 10 ATTRIBUTES OF A DENTIST 25

CHAPTER 3: FACTORS 21–30
TOP 10 CHARACTERISTICS OF THE DENTAL OFFICE 35

CHAPTER 4: FACTORS 31–40
TOP 10 ASPECTS TO LOOK FOR IN A DENTAL TEAM 49

CHAPTER 5: FACTORS 41–50
TOP 10 PIECES OF TECHNOLOGY YOUR DENTIST
SHOULD HAVE 63

CHAPTER 6: FACTORS 51–60
TOP 10 AMENITIES YOUR DENTIST SHOULD OFFER 81

CHAPTER 7: FACTORS 61–80
TOP 20 SERVICES YOUR DENTIST SHOULD PROVIDE 93

CHAPTER 8: FACTORS 81–90
TOP 10 THINGS A DOCTOR SHOULD DO BEFORE
STARTING COMPREHENSIVE DENTISTRY 123

CHAPTER 9: FACTORS 91–100
TOP 10 COMPONENTS OF YOUR DENTIST'S WEBSITE ... 133

CHAPTER 10: FACTOR 101
THE MOST IMPORTANT FACTOR 143

ACKNOWLEDGEMENTS 147

REFERENCES 151

INTRODUCTION

There are many factors to consider when making any decision. Many of you have been in this type of situation: an important evening is coming up, like a loved one's birthday, anniversary, graduation, or meeting is happening. You have to pick a great restaurant to go to. When deciding on a new restaurant, you will do a little research. You may look at the Zagat ratings, which give a nice description of the restaurant, along with ratings of food, décor, service, and cost. You may ask your friends or family about their experiences at their favorite restaurants. You may look at the restaurant's website. You may read a review from the local newspaper. You consider a number of factors: the type of food and drinks they serve, the ambiance, the availability, the location, the quality, the reviews, the reputation, and on and on. Whatever the case may be, you think about it before you decide. Going to a restaurant is usually a fun experience, and if it exceeds your expectations, you will probably go back a number of times. You will also likely tell other people about it. I hope that when you find the right dentist after having read this book, it will have the same result.

Dentistry is a profession that has been around since at least 7000 BC: the first evidence of dental procedures was found in a Neolithic graveyard in western Pakistan.[1] Dentistry has evolved dramatically in the last 20 years, with the emergence of dental implants to replace missing teeth, the popularity of whitening[2] and porcelain in the new era of esthetic dentistry, and most recently, the introduction of laser dentistry and 3D digital dentistry. In this new age of dentistry, you want to know the following 101 factors to make an informed decision about your oral health and dental care.

Going to the dentist is an extremely important aspect of your health. When choosing your dentist, make sure they are aligned with your hopes and wants. Verify that you are being listened to, respected, and valued.

The goal of this book is to help you to understand the one-hundred-one factors that you should look for when you walk into a dentist's office that deserves to have you as a patient. My office's goal and my team's goal is to embody these 101 factors. We are a work in progress, and hope to achieve these factors to the best of our ability every day. Does it always happen? No, but we try our best every day, and when we realize we have not embodied all of the factors, we step back, look introspectively, and get it right.

Dentistry is a science that is constantly evolving with vast advancements and improvements over the last couple of decades. Over time, new factors will emerge, and many of these new factors will be written.

You can read this book cover to cover. Good for you if you can! You are well on your way to becoming a dental dork like me!

You can look at chapters that address certain topics and reference them, or look at individual factors. You have full control over how you choose the right dentist for you. In other words, you make the call!

You have just made an important step toward creating the smile of your dreams. By reading this book, you will gain some insight into the importance of oral health and understand what factors to consider when choosing the right dentist for you. There are hundreds of aspects to consider, and this book discusses 101 of them. Thus the name: *Dental 101*.

Who enjoys going to the dentist? You would be surprised. Once you find the right dentist, the right dental office with the right attitude, and the right fit for you, going to the dentist will be as pleasurable as going to your salon, your spa, your coffee house, or your favorite restaurant.

Some of you may think that going to the dentist is scary, but it no longer needs to be. Sedation and new pain-free techniques have made dentistry extremely gentle and pleasant. You have options, and there are thousands of dentists around the U.S. that will take amazing care of you.

The objective of this book is to help you understand dentistry, to educate you about oral health, to help you determine what you want to achieve, and to help you find what you are looking for in a dentist.

There are many choices available to people today. With the advent of the internet and Google, finding a dentist is as simple as a search for Dentist followed by your zip code. How do you know you will get what you are looking for? How do you get the service you deserve? How important is it to you to get the quality care you deserve? What are you looking for in a dentist?

These topics will be discussed at length. Before we begin, I must warn you: I am male, and I write from the male perspective. Just know that "he or she" and "his or her" in terms of the dentist make things difficult to read and write, so you will see "he" in reference to "the dentist" often. Female dentists are amazing caretakers, a huge asset to dentistry, and quickly making up more and more of the dental landscape. I apologize for the chauvinism.

With that said, let's start with the foundation: The reasons to go to the dentist.

CHAPTER 1
FACTORS 1–10
TOP 10 REASONS TO GO TO THE DENTIST

Factor #1: To Keep Your Teeth

Dental hygiene is crucial to good health. Our health depends on the food we eat, so we have to take particular care of our teeth. Teeth are primarily used to break food into small pieces, and then chew them until they are in a state fit for digestion. Of course good teeth also produce a sparkling smile! Our teeth are unique to us. An anthropologist would be able to tell us our age and what race we belonged to, from our teeth.

Human beings have 20 primary teeth (also known as milk teeth and deciduous teeth) and 32 permanent teeth. Proper care of teeth begins as soon as they first appear. In fact, the first visit to a dentist should take place no later than 6 months after the first tooth appears, and definitely before your child's first birthday.

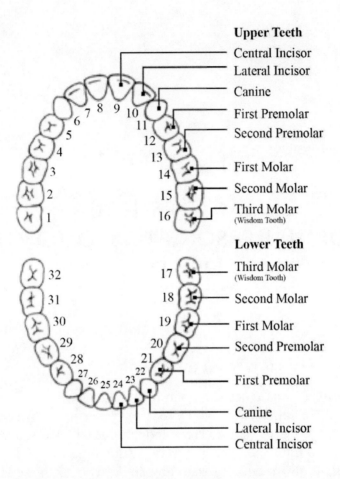

Upper Teeth

Central Incisor
Lateral Incisor
Canine
First Premolar
Second Premolar

First Molar
Second Molar
Third Molar
(Wisdom Tooth)

Lower Teeth

Third Molar
(Wisdom Tooth)
Second Molar
First Molar
Second Premolar

First Premolar

Canine
Lateral Incisor
Central Incisor

The structure of our teeth varies according to their position. The top front teeth are used for biting, while the teeth on the lower jaw are mainly used for breaking up and grinding the food. Further, teeth are divided into the incisors, canines, and molars. Teeth are made up of the enamel, which is the outermost layer. Underneath the enamel is dentin, a protective layer which also supports the crown. The central pulp is what contains the blood vessels and nerves. Below the gumline, there is cementum which helps the periodontal ligaments to attach the gums to the teeth. The alveolar bone is the jaw bone, which holds all the teeth in place. The gums, or gingiva, are tissues that cover the jaws. The roots of the teeth are normally found in the gingiva.[3]

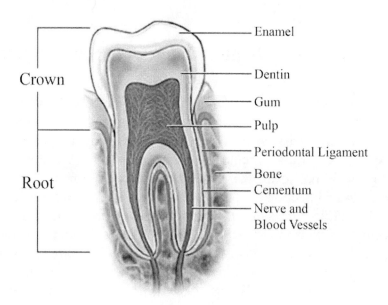

The dentist will give you the right advice on diet and home care of your teeth. Regular visits to the dentist will keep your teeth and gums healthy and strong, and they will allow the dentist and hygienist to make sure no incipient diseases or problems are present. There are literally millions of bacteria in the mouth. Some are not harmful, but some do attack the teeth and gums. The harmful bacteria are found in a sticky film called plaque, which collects on all of our teeth. This hardens over time to form tartar (also known as calculus).[4] Only a dentist or hygienist can remove this. Plaque leads to dental cavities or caries, a disease which damages the enamel, dentin, and cementum of our teeth. Plaque also leads to gingivitis and periodontitis which damages the gums and bone around our teeth. Over time, periodontitis and severe decay leads to tooth loss. Daily brushing, flossing, rinsing, and massaging of gums at home is essential, and significantly contributes to having great teeth and healthy gums. Going to the dentist will help you prevent periodontal disease and cavities, making it possible for you to keep your teeth longer.

Factor #2: To Maintain Excellent Overall Health

Oral hygiene is directly connected to our overall health. There are a large number of bacteria in our mouth, some harmful, and some not. When we regularly brush and floss our teeth, we ensure that our mouth is cleaned of all bacteria. Healthy gums prevent the harmful bacteria from entering the blood stream. When our oral hygiene is not ideal, the dangerous bacteria form plaque. This leads to an infection in the gums, known as periodontal disease or periodontitis. Once this happens, when we brush and floss our teeth, our gums start bleeding. This is a clear indication that our gums have been infected. Bad breath is another warning of oral ill-health. Tooth decay, gingivitis, and periodontitis are the first problems that occur. Worse, there is a strong chance that the bacteria will even get into the blood stream.

Gingivitis, the beginning stage of gum disease, is often not noticed by us. This is why visits to the dentist must be very regular. Gingivitis, undetected and untreated, will lead to periodontitis. Trouble in the mouth can be quite debilitating, and often very painful. Oral inflammation will consequently result in inflammation in other parts of the body. There is also evidence that oral inflammation is an indication of other health risks. Periodontal disease has been proven to be linked to coronary artery disease (leads to heart attacks), stroke, diabetes, complications during pregnancy (birth of pre-term and low-weight babies), and osteoporosis. Prolonged neglect of oral health can also lead to oral cancer.

For those that would like some evidence, read some of the studies that validate these claims in the following paragraphs. If you would like even more evidence, go to Google and search PubMed. Then type in any key words you want to know more about. PubMed is maintained by the National Institutes of Health (NIH) and the U.S. National Library of Medicine, and comprises more than 19 million citations for biomedical articles from Medline and life science journals. It's a great resource for any medical or dental questions and is based on scientific data from reputable sources, universities, and hospitals.

The link between gum disease and heart disease has been clearly proven. One of the most prominent studies comes from Dr. Moise Desvarieux, MD, PhD of Columbia University. In a study published in the February, 2005 issue of *Circulation: Journal of the American Heart Association*, he and his colleagues concluded that there is a direct relationship between periodontal disease and atherosclerosis (Atherosclerosis is defined "hardening of the arteries").[5] This study comes from a physician, not a dentist.

The correlation between gum disease and diabetes is proven by studies out of Stony Brook University. Dr. Maria E. Ryan, DDS, PhD has shown us in a study published in the October, 2003 *Journal of the American Dental Association* that "If there is oral infection and inflammation, as with any infection, it is much more difficult to control blood glucose levels".[6] Even the American Diabetes Association Vice President of Clinical Affairs, Dr. Sue Kirkman, MD, agrees. She stated "It is now clear that periodontal disease can make diabetes control worse, and there is even some evidence that it increases the risk for diabetes complications." One of these complications includes kidney failure. One study from NIH (National Institutes of Health) by Shultis proved that diabetics with severe periodontal disease had 5 times more likelihood of getting kidney failure than diabetics with no teeth (and no gum disease).[7]

Oral health has been proven to be correlated to pregnancy in a number of studies as well. One study out of Harvard was published in 2008 proving that periodontal disease is a risk factor for poor pregnancy outcomes in women.[8] Another study discussing this correlation is by Polyzos published in the *American Journal of Obstetrics and Gynecology* in the recent

March, 2009 issue. It concluded that periodontal therapy during pregnancy dramatically reduced preterm birth (PTB) and significantly reduced low birth weight (LBW) infant incidence.[9]

Osteoporosis has also been linked to periodontitis. One extremely relevant study out of Vanderbilt shows that osteoporosis, which affects over 10 million post-menopausal American women, involves the same messenger molecules called cytokines that are found in periodontitis.[10]

Knowing this, you need to be extremely vigilant about your oral health and hygiene. Brushing and flossing your teeth twice a day is the first step towards having clean, healthy teeth and gums. Using a fluoride toothpaste and mouthwash help. Proper nutrition and good eating habits contribute greatly to oral health. Dairy products, lean meat, fish, green vegetables, fruits, and whole grains, when included in the diet provide us with Vitamins B, C and D, as well as Calcium, Phosphorous, Iron, Magnesium, and Zinc, which keep our teeth and gums strong and healthy. Often just being aware of and bringing about lifestyle changes can lead to great oral health, and consequently overall health. In case braces or dentures have to be worn, careful and regular monitoring by the dentist will also contribute to overall health and well-being.

Factor #3: To Screen for Oral Cancer

One American dies of oral cancer every hour of every day.[11] Oral cancer claims over twice as many lives as cervical cancer.

Cancer is an uncontrolled growth of cells that form a tumor. Tumors that are not harmful do not invade the surrounding tissues. These are benign tumors, and they can be removed simply. Malignant tumors, though, enter the surrounding tissues and organs, damage them, and spread all over the body. These malignant tumors are often life-threatening.

Oral cancer may affect the tongue, floor of the mouth, back of the throat, lower lip, tonsils, or salivary glands. From here it can spread to other parts of the body through the lymph nodes (part of your immune system).

Causes of oral cancer[12] include:

- *Tobacco*—smoking, chewing, or dipping snuff are associated with 70–80% of all oral cancer cases.[13]
- *Alcohol*—heavy drinking, especially if combined with smoking.
- *Constant irritation*—from cracked teeth, failing fillings, or worn dentures.
- *Sun*—too much exposure to sunlight.
- *Poor dental hygiene and oral care*

Besides these, if you have a history of cancer of the neck and head, your susceptibility to oral cancer increases.

Symptoms to watch out for are:

- Red, white, or mixed red and white patches inside the mouth or on the lips
- A sore in the mouth or on the lip that is not healing
- Bleeding in the mouth
- Loose teeth
- Pain or difficulty while chewing and swallowing
- A lump in the neck
- A severe earache that is not healing
- Pain and difficulty while wearing dentures
- Thickening in the cheek
- Numbness in the mouth
- A feeling of soreness in the mouth
- A change in the voice

Oral cancer can be detected using visual oral cancer screening, dental x-rays, cone-beam computed tomography (CT) scan, or magnetic resonance imaging (MRI). Today, we can also detect very early signs of oral cancer using Velscope, a revolutionary method utilizing the same fluorescence that lung specialists use for cancer detection. We will discuss Velscope further in Factor #42.

If oral cancer is suspected, the dentist will make a recommendation based on the size, type, and location of the affected area. Often, the dentist will take a brush biopsy (a very simple procedure requiring no anesthesia) and send it to an oral pathology laboratory.[14] A preliminary diagnosis is made, and then the dentist and patient sit together to discuss the next step in treatment.

Approximately half of people with oral cancer will live more than five years after diagnosis and treatment. If the cancer is detected early, before it has spread to other tissues, the cure rate is nearly 75%. Unfortunately, more than half of oral cancers are advanced at the time the cancer is detected; most have spread to the throat or neck by the time they are diagnosed.[15] Early detection can improve the chance of successful treatment.

Oral cancer is aggressive, and the most effective way to help improve survival rates is early detection and treatment.[16] Oral cancer accounts for 8% of malignant types of cancers. Men have twice the chance of getting oral cancer than women do, especially men over age forty.[17] The American Cancer Society recommends oral cancer screening exams every three years for patients between the ages of twenty and forty, and then annually thereafter. During your next dental appointment, ask your dentist to perform a comprehensive oral cancer screening with Velscope.

Factor #4: To Prevent Gum Disease

In dental literature, gum disease is known as periodontal disease. *Periodontal* means *around the tooth*. Periodontal ligaments attach our gums firmly to our teeth so that no germs, or the toxins they produce, can get through.

Healthy gums are firm and look pink. As mentioned before, the mouth is full of bacteria, some of which may be harmful. These bacteria encase themselves in a sticky film called plaque. They permeate into our gums and the blood vessels within, and produce toxins.[18] In order to fight these toxins, our bodies cause the gums to get inflamed. This is when the gums start bleeding, particularly when we brush our teeth or when we eat. This is a condition called gingivitis.[19]

If left untreated, the plaque hardens to form tartar. The germs spread deeper into the bone, and the condition is then called periodontitis.[20] If nothing is done at this stage, the periodontitis (also called periodontal disease; it was referred to as *pyorrhea* in literature prior to 1945) will progress to moderate periodontitis and then advanced periodontitis. Eventually advanced periodontitis causes teeth to become loose and fall out. The most surprising thing about periodontitis is that it causes no pain. That is why most of us tend to ignore it. Besides tooth loss, periodontitis has been linked to many systemic conditions, including heart disease, stroke, diabetes, and osteoporosis.

The best ways of preventing periodontal disease are:

- *Regular Dental Examinations*—This should be done at least once or twice a year. X-rays of the mouth might need to be taken every year. The dentist or hygienist will do a periodontal screening and periodontal charting to measure the gum pockets to evaluate the health of the teeth and gums.
- *Healthy dietary habits*—Plan a well-balanced diet. Reduce, and if possible, avoid sugar. In case we have had something sweet to eat, we need to brush our teeth soon after.
- *Giving up tobacco*—especially smoking. Tobacco has been shown to accelerate periodontitis.[21]
- *Great daily oral hygiene*—Brushing, flossing and gargling with a good mouthwash are the basics. Follow your dentist's instructions on the best method of brushing our teeth as well as the kind of toothbrush, toothpaste, and mouthwash that should be used. The tongue needs gentle brushing as well. It is also a good idea to learn the correct way of flossing your teeth. In case we are in a situation

that does not allow for brushing after our meals, rinsing the mouth with water is a must.

Periodontitis is a disease that is progressive and persistent. If you want to prevent periodontitis or stop it in its tracks, regular dental visits and cleanings are a necessity.

Factor #5: To Have a Bright White Smile

A smile brightens up not only our day, but also evokes a smile from those we meet. A smile has a remarkable impact on perceptions of one's attractiveness and one's personality. Previous research in the field of psychology has shown that attractive people are perceived by others as more successful, intelligent, and friendly.[22] Teeth alone can have an impact on overall attractiveness and perceptions of personality attributes[16]. One's smile clearly plays a significant role in the perception that others have of our appearance and our personality. Self-esteem and self-perception has also been proven to be positively affected by an attractive smile.[23]

Teeth get discolored because of:

- *Age*—as we grow older, there is a build-up of plaque. Dentin, which is the layer below the enamel gets darker, and causes the teeth to get discolored.
- *Food and beverages*—eating foods that have food coloring, and drinking beverages with caffeine, stain the teeth over time.
- *Tobacco*—smoking, or chewing tobacco cause discoloration.
- *Gum disease and teeth decay*—if we neglect our teeth, this will discolor our teeth.

The dentist would first treat any dental problems, and then offer cosmetic solutions. We need to discuss our options with him so that we are clear about the procedure of stain removal as well as the cost.

The various chemical whitening options are:

- *Hydrogen peroxide or baking soda*—if there are only a few stains, then the teeth whitening products we need are available over the counter. These are hydrogen peroxide or baking soda. Covering the enamel with these does the trick.

- *Whitening strips and whiteners*—these are not permanent whitening agents, but will suffice for a short while. We need to check with the dentist before using these because they may adversely affect and destroy the enamel of our teeth. Gels contain glycerin which draws out the moisture from the enamel. Another point to remember is that whitening strips will not whiten all the teeth. They can be used only on 6 of the front teeth.

- *Laser whitening*—the whitening gel contains peroxide. This is put on the teeth, and then laser beams are scanned across the teeth. These beams activate the gel and the stain is removed. This method is painless and very effective. Each session lasts for an hour or an hour and a half, depending on how bad the stain is and how white we want our teeth to be.

- *Bleach*—should you decide on bleaching agents, the dentist will give you a bleaching kit. He decides the percentage of whitening agent to be used. He will also see to it that the mouth tray is custom-fitted so that you get a better seal. The whitening gel is put into the tray, and when you seat it, the whitening gel spreads evenly over the teeth.

Another definitive, permanent option to whiten teeth utilizes masking technique: *veneers* or *crowns*.

- *Veneers or Crowns*—A veneer and a crown share a number of common characteristics. Both are fabricated in a dental laboratory and enable patients to achieve white, aligned teeth. Both are custom-tailored to the patient's mouth. Veneers are always made of porcelain. Crowns today usually have porcelain on the outside, but can sometimes be made of metal (hopefully not in the front of the mouth). The biggest difference between a veneer and a crown is that a dental crown encompasses the entire tooth, whereas a veneer only encases the front part of the tooth. A veneer covers the part of the tooth that is visible when a person smiles. Dental crowns are used to repair teeth that are cracked, fractured, decayed, or previously filled with large fillings. Dental crowns are stronger than dental veneers, but do require removal of more tooth structure. Veneers are used on teeth that are still strong and healthy, even though they may be discolored, have gaps, or crooked. Veneers only cover the front part of the tooth, so the tooth does not need to be reduced as much as a tooth needing a dental crown. Veneers are strong but brittle. Certain factors also need to be considered when planning a veneer or a crown, including history of decay, whether or not someone grinds or clenches their teeth, and gum disease.

Factor #6: To Prevent Cavities

There are many bacteria in the mouth. If we do not brush our teeth or rinse our mouth after we eat, these bacteria, the *Streptococcus mutans* and other anaerobes, work on the remaining food particles to produce an acid which corrodes the enamel of our teeth.[24] Gradually, a cavity is formed. Your teeth have some ability to make repairs using the minerals that are in the saliva. However, the acid is extremely corrosive, the destruction is quite rapid, and the process is initially painless. It is only when the cavity is formed that we experience bad breath and pain. If problems of teeth and gums are genetic, then we need to be doubly careful about our teeth.

Dental cavity prevention requires the following:

- *Regular brushing of teeth*—at least two minutes, twice every day. Check to see if the toothpaste has fluoride, which hardens the enamel and prevents the bacteria from eating into the teeth. Get a new toothbrush every 3 months.
- *Daily flossing*—crucial to preventing cavities between teeth, and to ensure that no food particles remain in the mouth.
- *Use a mouthwash*—an anti-cavity fluoride rinse strengthens enamel on teeth. Try to utilize an alcohol-free rinse to help protect the roots of adult teeth, which become vulnerable as gums begin to recede with age.
- *Maintaining a healthy diet*—including whole grains, dairy products (source of Calcium), fruits, and vegetables (for Vitamins A and C). Drink at least 8 glasses of water per day. Cut back on artificial fruit juices, soda, and power drinks as these weaken the enamel due to acid and sugar. Gradually cut down completely on sugar, especially high-fructose corn syrup, because these aid the bacterial activity. Lessen the amount of starch as well, since starch, when broken down, forms sugar.
- *Drinking tea*—both green and black tea without sweeteners prevents excessive plaque from forming. Drink water afterwards, because tea does stain teeth.
- *Chewing gum*—sugarless gum or chewing gum which has Xylitol keeps the saliva flowing.[25] The alkalinity of saliva neutralizes the acid from the bacteria, as well as flushes out the debris of food particles. Be aware that chewing should not exceed a few minutes a day, because excessive

gum chewing will lead to TMJ (your jaw joint in front of your ears) problems.

- *Dental sealants*—are protective coatings that can be applied on biting surfaces of the back teeth.
- *Regular visits to the dentist*—are imperative to detection, prevention, and treatment of cavities.

Factor #7: To Prevent Bad Breath

Bad breath, or halitosis, could be the result of either poor oral hygiene or problems in the gastrointestinal tract. At the root of both are bacteria. And so, first we need to find out why we have halitosis, and then go about treating it. The most common cause of bad breath is periodontitis. In one recent study, 76% of people got bad breath from either periodontitis or gingivitis.[26] Some guidelines to prevent bad breath are:

- *Brushing and flossing*—this needs to be done at least twice a day. We need to clean the tongue as well at this time with a brush, or a tongue scraper. After brushing, flossing, and cleaning the tongue, rinse the mouth with a mouthwash.
- *Drinking water*—the minimum amount of water is at least 8 glasses a day. The reason is that the anaerobic bacteria form volatile sulfur compounds (VSCs) from protein and sugar. This causes the bad smell. Water dilutes the VSCs, and also inhibits the bacteria from producing this. Xerostomia (or dry mouth) makes it easy for bacteria to multiply and stick around, so it's really important to drink water throughout the day.
- *Drinking tea*—both green and black tea are good since they prevent the growth of bacteria.
- *Consuming certain foods*—including apples, celery, cucumbers, and carrots. Crunchy fruits and vegetables keep the mouth clean, and produce more saliva. Chewing

on spearmint, coriander, parsley, cardamom, tarragon, basil and rosemary or drinking infusions of these, prevent halitosis and make for excellent digestives as well. Yogurt is wonderful for reducing plaque and gum diseases. Dairy products like cheese and milk supply Vitamin D. Citrus fruits, melons and berries are a good source of Vitamin C. Vitamins inhibit bacterial growth. If your diet includes garlic and onions, ensure that they are used in moderation and that you wash your mouth after eating foods containing these.

- *Having breakfast*—skipping breakfast causes the stomach to produce acids causing bad breath.
- *Stop smoking*
- *Visiting the dentist regularly*—this is the most important guideline. A dentist will be able to diagnose any oral causes of halitosis and prevent them from developing.

Factor #8: Help You Avoid a Dental Emergency

You know you have a dental emergency when:

- You have severe pain in your mouth, especially if it lasts longer than an hour
- You are bleeding profusely from your mouth
- You have suffered a facial injury
- There is a swelling in your mouth or face
- There is a swelling in your gums
- A toothache that is waking you up

The best ways of avoiding any kind of dental emergency are by:

- Ensuring that you have excellent oral hygiene, so that your teeth and gums are strong and free of disease.
- Following all of the dentist's instructions regarding brushing, flossing, cleaning the tongue and using the right

kind of mouthwash, as well as choosing the right kind of toothbrush, toothpaste, tongue cleaner, and mouthwash.

- Avoiding eating hard foods such as hard candy and corn kernels that have not popped. Also avoid chewing ice.
- Never using your teeth to cut strings, open packages, hold metallic objects like nails or screws, or tear anything.
- Using a mouth guard when playing contact sports like baseball, football, basketball, hockey, power lifting, MMA, and boxing.
- Regularly visiting the dentist. The best way of avoiding any kind of dental emergency is to keep your regular appointments at your dentist's office. There is a two-fold benefit—information about your teeth, what to do in an emergency, and even how to prevent a dental emergency; and to understand your options in case an emergency comes up. For instance, if your dentist identifies any problems with your teeth and gums, you need to get those issues addressed as soon as possible. Delaying or putting off dental work that needs to be done might have serious repercussions. You also need to understand the costs involved, so that you can plan out a timeline, and also consider third party financing.

You must know that accidents do happen in spite of the best dental care. A dental emergency could be a toothache, a broken tooth, a tooth that has been knocked out, a broken or lost crown, possibly a broken jaw, an injury on the tongue or lip, a broken denture, an abscess in the gums, an infected wisdom tooth, a filling that has failed, or it could be something which has gotten caught between the teeth.[27]

Factor #9: To Establish Dental Goals

It is extremely important to have clear dental goals. This enables a dentist to maintain a record of not only your teeth and dental health but also of your general health, since one impacts the other.[28]

Some dental goals are:

- Proper long-term health of our teeth
- Excellent oral hygiene
- Keeping track of any pain or feeling of discomfort in the mouth. Then having the problem, incipient or obvious, diagnosed and treated immediately.
- Healthy gums—If the formation of plaque is checked in the early stages, then you can rule out getting gingivitis or periodontitis
- Timely correction of any disorder, such as caries, gingivitis, periodontitis, oral cancer, replacing missing teeth, removing wisdom teeth, or restoring teeth to full function
- To avoid losing any teeth
- To prevent oral cancer
- To consult with the dentist in the event of other medical problems in order to see if the teeth and gums have been affected
- To have a beautiful, white, sparkling smile
- Getting implants, bridges or dentures done if required

A treatment plan is necessary in the event of any disease. This is to eliminate disease, and restore the teeth so that they can function normally. The treatment plan, worked out with the dentist, has the following steps:

- *Developing the plan*—done after a consultation, x-rays, photos, models, a comprehensive dental exam, and periodontal charting.
- *Plan sequencing*—the entire plan is worked out step by step.
- *Plan presentation and informed consent*—the patient knows exactly what is going to happen since the dentist

suggests a variety of techniques, procedures, and products that are possible, and then decides on the best course.

- *Comprehensive plan execution*—depending on the general health and costs involved, it may be necessary to phase the treatment.
- *Plan modification*—changes in the condition of the patient as well as access to advanced technology will require corrective measures to be taken.

Factor #10: You Deserve It!

Your smile and your teeth are important to your self-esteem[17]. Strong teeth, healthy gums, and clean mouth are the foundation to an amazing smile. To achieve this, you need to take good care of your teeth and gums. Regular dental visits must become a part of your life. Increasingly it is clear that your body works as a whole, especially in terms of the connection between our oral health and systemic health[22].

Also, more and more it is becoming evident that lifestyle dictates what kind of health we have. We need to examine our lifestyle and do what it takes to ensure that we do not do anything to harm ourselves.

The first thing we need to scrutinize is our diet. A balanced diet not only guarantees good health, but also safeguards our teeth and gums. Eliminating or at least cutting down on fruit juices, energy drinks, soda, caffeinated beverages, sweet or starchy foods (since starchy foods break down to sugar) is good for sound health and strong teeth. A nutritious, balanced diet is aided by a sound dental care regimen.[29] This involves meticulous daily care of the teeth and gums. If we feel that we deserve a prettier smile, then your dentist will recommend cosmetic procedures. He will tell us the various techniques and procedures that would make sure that you look and feel better. Whitening of the teeth, shaping the teeth, dental implants, dental bridges, dental crowns, gum procedures, and braces are all methods you could use to have a gorgeous set of teeth. Sometimes a combination of methods is used to get the effect that you are looking for. And why not?

You deserve to look and feel great! All you have to do is to work out all the details so that the dentist is clear about what you want, and you are clear about what is involved, including the cost. If the treatment is something that you have set your heart on, but is too expensive, then the dental team will tell you what you can do in terms of getting insurance or third party financing, or going in for a phased treatment. You deserve to get the treatment you want!

CHAPTER 2
FACTORS 11–20
TOP 10 ATTRIBUTES OF A DENTIST

Factor #11: Competent

A dentist's office is not usually a place where people go happily. A competent dentist automatically inspires confidence and this in turn removes the lurking fear a visit to the dentist causes. Receiving a degree in dentistry is the basic qualification that entitles a person to become a member of the profession. He would have acquired a body of knowledge as well as the necessary skills. The dentist would then decide if he wants to be a specialist, and will work towards achieving this goal. Along the way, the dentist would develop the skill of diagnosing what is wrong, explaining it to the patient, and deciding on the necessary course of action. A very important characteristic of competency is to be able to identify with the patient. Compassion for the patient's distress arises out of knowledge not only of dentistry, but also of psychology.[30]

A competent dentist would never rest on his oars, but constantly seek to keep in touch with the latest developments in the field of dentistry. New knowledge and new methods and procedures are constantly being discovered and developed. Keeping abreast of these developments through continuing education increases the dentist's confidence and competency. This is immediately visible in his work and in the way he handles his patients. A competent dentist would collaborate with other dentists through study clubs to cultivate dialogue and discuss dental literature. The competent dentist would interact with other health care professionals through correspondence, since the human body requires holistic treatment. Competency in a dentist would be evident with his basic life support (BLS) systems. In the event of an emergency in the office, you want a dentist who is up to date with his BLS.

Good communication skills are an absolute necessity to competency, as well as good listening skills. If a dentist is empathetic and has respect for human dignity, he will be able to establish a rapport, such that the patient gives all the information required fearlessly. The dentist should also be aware of and sensitive to cultural differences. The ability to create the right atmosphere, ask the right questions, and evaluate the data received in order to make an informed opinion and take the right decision depends on good communication skills. Organized and systematic methods of thinking, listening, and functioning are hallmarks of a competent dentist.

A competent dentist would not only be of an intellectually superior mind, but would also be physically, mentally and emotionally thoroughly fit. How comprehensive his diagnosis and treatment plan are depend entirely on his competency. Competency includes integrity. Competency involves working out systems that make it possible to be accessible at all times.

Factor #12: Trustworthy

If we think about it, the dentist holds a very important place in your lives. There is archaeological evidence to show that dentists have been important care-giving members of society from

3000 BC.[31] The mouth is a very sensitive area, and you need to know that the dentist you go to is knowledgeable, passionate, honest, and responsible. Professional qualifications include a thorough knowledge of the subject and skills required.

A trustworthy dentist will treat you as unique individuals, and not divulge your particular case history or your confidences to any one at any time. He is sincere and competent, and consequently will explain the diagnosis and the treatment plan truthfully as well as gently. He is a person whom you can trust with the extremely sensitive issue of health history. We are confident that, should the need arise, he will give us the right kind of financial advice.

A trustworthy dentist will deliver what he has promised. Should a second opinion be required, an honest dentist will make the right recommendation. Being trustworthy involves a responsibility in being up-to-date in the field of dentistry. This means regular continuing education and study clubs. This automatically translates to sincerity and you should immediately sense the truthfulness of the dentist, especially if changes need to be made in the treatment plan.

Since the dentist is dealing with such a sensitive part of your body, it is imperative that he use quality equipment and quality materials. This means that he uses the best dental laboratory available to him. In the light of the fact that he is aware of the latest developments and techniques that can be used, he is not afraid to use these to advantage. A trustworthy dentist will use the same high quality of service regardless of any patient's financial status.

The question that arises is how you can find such a dentist. Usually, it is by word of mouth, or by personal referrals. Often, we can find the dentist by the yellow pages, Google, Yahoo, or your local Suburban Woman's newspaper.

In order to investigate the dentist, you could simply call the office and evaluate the service, and likely make an appointment for a cleaning. When you arrive at the dentist's office, you will be able to get a feel of the office and the team there. Talking to other patients and even the other team members there will help you understand the set-up and systems the dentist has. Finally, from the way the doctor receives you, his questions to you, the way he listens to you[32] and your dental condition, the way his hygienist goes about cleaning your teeth, and his diagnosis about the general condition of our teeth and gums, all will indicate to us whether we can trust the dentist or not.

Factor #13: Thorough

A true professional is extremely thorough in all that he does, and a dentist is no exception. Dental problems could be due to a problem in another part of the body, or they could create a problem elsewhere in the body. That is why it is important not to neglect either any pain or discomfort that might occur in our mouth. The initial examination that a thorough dentist would do would be very detailed and comprehensive, in that he would get to know your oral condition in its totality. His questions about your oral condition and health history also give you an indication of his methodology as well as his philosophy.[33] His tentative diagnosis would be supported by a thorough clinical examination which would include:

- A careful exam of all our teeth to see if there is any decay, as well as if there is any build-up of plaque. Also, if we have had any restorations done, he will check their present condition. He would further recommend a full set of dental x-rays of all the teeth or a cone-beam CT scan to evaluate the mouth in three dimensions.
- A check-up of the gums using a periodontal measurement probe. In case the gums have receded, possible reasons and solutions for this can be discussed.
- An evaluation of the *bite*. This shows if there might be excessive wear of some teeth, and also the balance of the

teeth. If there is any muscle pain, or sensitivity, it becomes evident at this point.

- A check for oral cancer. If the dentist is well up on the latest developments and techniques used in dentistry, he will have Velscope in his supply of tools. This is used to examine the mouth for oral cancer, as they detect any abnormal growth of cells and tissues in the mouth.

The thorough dentist, thus, not only diagnoses the immediate problem, but also might ask for more investigations to be done in order to address your overall health. Once all the results come in, the conscientious and thorough dentist would review the findings with you so that you understand the present condition, as well as what needs to be done to correct the problem areas. The thorough dentist would then see the treatment plan through right to the finish.

Factor #14: Punctual

Going to any doctor can be stressful. The last thing you want to add to your stress is a long wait. Your time is important and valuable. Keeping and maintaining a good schedule is paramount to helping you have a great experience with your dentist.

A dentist that is always late indicates a number of potential issues. It means that the systems are not well established in the office, that your time is not as important as his, or he is unable to do certain things in an allotted amount of time. Exceptions certainly do exist, including emergencies, complications in procedures, and a late patient before your appointment.

In the event of a tardy dentist appointment, you certainly deserve an apology. If you do get one, accept it, and if you feel like you will not have enough time for your appointment, reschedule. The doctor's office will definitely understand.

Factor #15: Jovial

A dentist does not have the most pleasant job with saliva and tooth debris getting shot into his face all day, but a dedicated dentist who loves what he does brings joy and energy into the office with him every day. One very important characteristic of a dentist is his smile. If his smile is genuine, it immediately inspires confidence and a sense of calmness in you. The dentist is actually an extremely important part of your life.

The personality of the dentist plays an important part to your overall experience. If he is warm and pleasant, that is the atmosphere he is going to create in his procedures. Pain in the mouth is not fun, but if the dentist can put you at ease, the rest of the time spent in the dental chair becomes manageable. The sincerity of a dentist is apparent from the way he asks questions about your issues and problems. Although he is deadly serious about his work, he gives an air of lightness to your procedure. That proves that the dentist loves what he does. This translates into every area of his work. Communication is very important, we know, and a jovial dentist discusses the diagnosis he has arrived at as well as outlines the treatment plan in a very nice and easy way.

A dentist who is jovial is very serious about his patients and in the care he gives them. He manages to extract our entire medical history in a manner that is sensitive to your feelings. A jovial dentist discusses the dental issues gently and pleasantly in a way that the patient is neither stressed nor frightened, and yet gets to know what the problem is, what needs to be done, and what the alternatives are. A jovial dentist keeps a fun atmosphere and never makes the patient feel guilty about their conditions, nor does he ever judge a patient. A jovial dentist is always accessible, and he is happy to help you even when he is on vacation.

Factor #16: Skilled

A skilled dentist is one who has the mental and physical capacity to look at you and come up with multiple ways to create solutions for your dental needs and wants. Skill comes in the form of diagnosis, prevention, treatment, follow-up, and maintenance.

When you go to a dentist, you expect a certain level of skill from him. All dentists in America have gone through at least 4 years of dental school after college. They definitely have the skill to do certain procedures. If you hope for someone who will provide you with comprehensive care that goes beyond traditional fillings and cleanings, look for a dentist who is committed to continuing education. These dentists have been extensively trained on more advanced procedures, such as laser dentistry, CAD/CAM technology, implant dentistry, TMJ treatment, full mouth rehabilitation, and cosmetic dentistry.

Make sure your dentist is committed to continuing education. Some prestigious post-doctoral educational continuums to ask about are Implant Seminars, Dawson, Pankey, SOBE, Misch, Hornbrook, Kois, and Gordon Christensen.

Factor #17: Friendly

Do you remember the story of *Snow White and the Seven Dwarfs*? Those seven dwarfs all had clearly different personalities, defined by their names. Who did you like the most? Some people are drawn to Dopey, some to Happy, and others to Bashful.

When looking for a dentist, look for the dwarf you want. You want to find a partner in evaluating, creating, and maintaining the smile of your dreams. This means you will be spending some time with that dentist, and you want to ensure that your time is spent in a nice atmosphere. A Happy, friendly dentist will make that time pass a lot more pleasantly than a Grumpy dentist. Grumpy was not the most popular of the 7 dwarfs.

Your dentist should care about your teeth and gums, but also care about you. If he asks you for your name umpteen times or doesn't talk to you at all, chances are that he isn't very friendly.

Factor #18: Attentive

A great dentist is very attentive in his work. Attention to detail when diagnosing ensures the dentist looks at all of the data with accuracy. Attention to detail when treating disease enables a dentist to remove any future problems. Attention to you as a person and as a patient gives the dentist the ability to treat you the way you want, the way you deserve.

Listening is a step that most dentists fail to take the time to do. Attentiveness when you are speaking is vital to getting the dental care you want. Make sure your dentist is attentive when you are talking to him.

Factor #19: Effective

An effective dentist is one who manages his team well and effectively communicates with patients. He allows people in his office to work independently but steps in when necessary. He treats his team with respect, and tells them their responsibilities professionally.

He can explain things in a way that makes sense to patients (who aren't fluid in the unofficial language of dental terminology). Many dentists assume too much about their patients, make pre-conceived judgments about them, or think that their patients don't need to be told any details. An effective dentist finds out what the patient wants to know and gives it to them.

Factor #20: Painless

In today's day and age, painless dentistry is truly available. One of the most common fears of going to the dentist is getting an injection.[34] New advances in dentistry include PFG and Profound, which is a topical anesthetic gel made from a powerful combination of three kinds of anesthetic: tetracaine, lidocaine, and prilocaine.[35] These three anesthetics literally remove the "prick" feeling people get when they receive an injection.

The local anesthetics that are available to us and the techniques dentists have to administer them enable you to feel absolutely no pain during a dental treatment. If you ever feel pain in the dental chair, tell your dentist. He will be happy to relieve the pain with some of these techniques.

CHAPTER 3

FACTORS 21–30
TOP 10 CHARACTERISTICS OF THE
DENTAL OFFICE

Factor #21: Clean

Cleanliness in a dental office certainly includes a clean environment. It also includes the personal cleanliness of all who work there. Any dental office must have a high standard of cleanliness. This prevents the transmission of diseases to both you and the dental care providers.[36]

There is an idiom that says "Cleanliness is next to Godliness". Cleanliness and sanitization of the dental office is the responsibility of the dental assistant and the dental hygienist, but ultimately needs to be monitored by the dentist. OSHA (Occupational Safety and Health Administration) mandates that after each patient, a complete wipe down of all surfaces is mandatory. The dental chair and other surfaces need to be wiped with a disinfectant. The headrest cover and patient bib need to be changed after every patient. All dental instruments used have to be cleaned, disinfected, and sterilized in an autoclave. Needles should always be discarded and other disposable instruments need to be put in a separate trash container, including suction tubes. A carpet, and a sofa set are repositories of dust and should be vacuumed regularly. All surfaces, including mirrors, need to be cleaned preferably with an antiseptic lotion, or a disinfectant spray. Towels and other materials that the dentist uses have to be changed frequently.

The water ejector, light handles, or any equipment including the x-ray equipment that is used, need a germicidal wipe. Water should be allowed to flow for a few minutes through the hand piece after it has been used. Suction hoses need to be cleaned with disinfectant spray.

Toilets and sinks should be clean. All books and magazines, furniture and electronic equipment need regular dusting and cleaning. Light and fan fixtures, air conditioners, walls, ceiling, windows, and doors have to be kept clean. Flowers and plants add life to a room, but these need care and attention. Garbage has to be removed every day, but the garbage after any dental surgery will need to be cleaned out after every patient.

When cleaning dental instruments it is essential that the face is covered and gloves are worn. Instruments are cleaned with ultrasonic solution in the ultrasonic machine. A noncorrosive enzymatic cleaner is a good cleaning agent. The machine has to run for the required time. Plastic instruments are cleaned with cold sterilization solution, usually glutaraldehyde. This agent, glutaraldehyde has been shown to be effective at killing the HIV virus.[37] Metal instruments must be sterilized in an autoclave, which uses high pressure steam and heat.[38] After cleaning the instruments they will need to dry. All cleaning tools have to be kept clean and decontaminated. It is best to replace them as often as possible.

The cleanliness of a dental office is the litmus test of a dental office's level of care. If you enter a dental office and get a feeling it isn't clean, turn around and find one that is.

Factor #22: Comfortable

A good dentist is essentially a sensitive person. He is aware that a visit to the dentist can be rather daunting and frightening. Ideally, visits to the dentist should start when the first tooth appears. Then, through the years a relationship or trust builds up between the dentist and the patient. In these cases, going to the dentist is not traumatic. However, the majority of people go to a dentist in the event of some dental emergency. When the dental office is comfortable, relaxing, and soothing, your fears are put to rest. Therefore, the décor and feel of a dental office requires a great deal of thought, care, and attention to the smallest detail.

A dentist has to be highly qualified and have state-of-the-art equipment to provide a high level of care. However, it is the way he treats patients that will make him the dentist you would choose to go to.

The comfort factor should start when you enter the reception and welcome area of a dental office. If all those who work in the dental office are friendly and well-trained, the first step towards comfort should be a warm welcome. This should be followed by some sort of a beverage bar with coffee, water, and drinks to help you relax before your appointment. If it is your first time to an office, the team should give you a tour, describing the office and its amenities.

Next is the appearance of the office. A clean, bright atmosphere, the right mix of colors, the right kind of lighting, and comfortable furniture increases the comfort factor. Spacious rooms are psychologically comforting. Specially and carefully selected pleasant music is soothing. The chair on which the patient sits should be relaxing. Knowing exactly what is going to happen makes the patient psychologically comfortable.

Modern dental offices have flat-screen televisions displaying dental information, and music piped throughout the office. There are headphones for you if the sound of the drill bothers you.

Once seated, expect to be pampered a bit. You should get some sort of a soothing neck wrap, napkin to keep you dry, and an eye-pillow or goggles to protect your eyes. After the clinical examination, you can listen to the music of your choice during your procedure. A blanket is provided if you are cold. The cleanliness of the office also goes towards the feeling of comfort that you are in good hands.

A major comfort factor is whether the dental work is painless. To this end, dentists use PGF to numb the area before getting anesthetic, perhaps Nitrous Oxide to relax and soothe you if anxiety levels are high, and perhaps sedation if you are extremely nervous.

After the dental work is over, you will be given a hot towel to clean off and a bowl of ice cream to remove the dental taste and help to reduce any swelling.

You should be able to drive home or go back to work unless you have been deeply sedated. If a surgery is completed, the dentist will prescribe antibiotics and pain medications as well.

Comfort is essential to a great experience at a dental office. It encompasses many elements, and it should be a factor when you choose the right dentist for you!

Factor #23: Welcoming

This is the very basic requirement in a dental office. Since visiting the dentist is not the easiest of tasks, it is important that the office be warm and welcoming. The dental office needs to be in an area that is accessible, and where parking is not a problem. The layout of the dental office is important too. Some offices are in medical-professional buildings, others in stand-alone buildings, and some are in strip malls. Since a dentist handles many personality types, it is important to keep the comfort and well-being of all his patients in mind. The welcome area is where stress levels might rise. Consequently this is a place which has to be done up with maximum care.

Pictures, flowers, soft piped music, educational video[39], books, and magazines will go a long way in pleasantly engaging the minds of the patients. The staff is attentive and trained to handle all kinds of patients and all kinds of situations. A smiling, considerate office administrator would handle all appointments efficiently besides, of course, welcoming patients. Dental assistants would be available in case patients just needed to talk to someone in order to overcome their fears, while waiting to see the dentist.

Tea, coffee or other light beverages and something to munch are always welcome. The dentist himself would be very welcoming when it was your turn to see him. If the dentist's office is done up with state-of-the-art equipment and with facilities to relax the patient, you feel welcome, and consequently, immediately relaxed. A dentist who has sound knowledge of his subject as well as of human nature realizes that each case is unique, and he will treat each patient accordingly, and with undivided attention. Remembering his patients and referring to their last visit immediately makes them feel welcome. The patient is treated with great care and gentleness by the whole dental team. Each patient is taken on a one-on-one basis and no one is kept waiting. Thus, whether it is a routine check-up for the whole family, or some dental emergency, all patients are sure that they will be welcome, listened to, and given the best possible treatment.

Factor #24: Modern

A bright smile, aligned white and sparkling teeth, a clean mouth, and healthy gums are required for a high sense of self-esteem and self-confidence. A modern dentist treats his patients in a holistic manner. He realizes that oral problems do not happen in isolation. Assuming that he really loves his profession and is committed to it, the first thing that will strike a patient is his dentist's knowledge and sincerity of purpose. The decor of the rooms allows you to be psychologically comfortable before an examination or a dental procedure. A modern dentist realizes that basic qualifications in dentistry are not enough. He has to know about and be familiar with all the latest developments and techniques through extensive continuing education. Modern dentistry includes implant dentistry, cosmetic dentistry, reconstructive restorative dentistry, digital dentistry, and laser dentistry. His office will have the latest equipment, and his methods of clinical examination will be modern and progressive.

What every patient wants is to know exactly what the issues are. Modern digital technology including cone-beam CT scanning, digital x-rays, and digital photography enables patients to see the condition of their teeth on large monitors. Intraoral cameras and digital x-rays help the doctor explain the diagnosis as well as the treatment plan. Cone-beam CT scanning allows patients to see every tooth, the jaws, and the joints in three dimensions.[40]

Preventive oral care using modern methods is what the dentist seeks to educate his patients in. The areas that the modern dentist would deal with are: general dentistry; porcelain crowns and veneers to straighten teeth, close gaps, and improve the shape and color of the teeth; whitening teeth using the latest techniques; placing and restoring dental implants; and laser technology to remove gum disease and remove the causes of it.

Pain definitely frightens patients, but modern dental techniques essentially eliminate this aspect. Various methods of eliminating pain depend on the patient's threshold of pain. The dentist is sensitive to this, and he could offer various options depending on his knowledge of the patient and the dental work that needs to be done. Removal or minimizing pain could be by sedation of the patient the night before a dental surgery, or by using nitrous oxide to relax and soothe the patient. The dentist might use a triple local anesthetic gel, or intravenous sedation.

Factor #25: State-of-the-Art

A modern dentist will have state-of-the-art equipment in his office because this is the only way he will be able to provide optimal care to his patients. People are not willing to settle for anything less than the best. Unfortunately, the dentist's office used to look like a torture chamber till not very long ago. This, added to the pain involved in any oral problem was enough to make a person visiting the dentist thoroughly scared.

Fortunately, though, those days are over. Dentistry includes so much more today from preventive dentistry to restorative dentistry to cosmetic dentistry to digital dentistry to sedation dentistry. Modern, progressive dentists are specialists not only in their own fields but have a basic knowledge of other medical areas as well. To that end, their offices are equipped to deal with most medical emergencies that might arise during a dental procedure. Many pieces of equipment make up a dental office. All dentists have a dental chair, hand instruments, and drills. State-of-the-art equipment would include:

- *i-CAT Cone Beam CT Scanner*: allows dentists to see teeth, jaws, and joints in three dimensions
- *Velscope*: helps in the early detection of oral cancer
- *Digital intraoral photography*: allows you to see what the dentist sees and allows the dentist to study your mouth subjectively
- *Digital low-radiation x-ray*: enables the dentist to get data on your teeth that is not possible with a visual examination
- *Whitening laser*: gives patients a whiter smile in about an hour
- *Soft tissue diode laser*: non-surgically reverses the causes of gum disease[41], allows the dentist to do various gum procedures[42] with little to no pain later
- *Ultrasonic machine and autoclave*: to disinfect and sterilize instruments[43]

- *Flat-screen monitors*: to allow you and the dentist to see x-rays, photos, and charting
- *Latest software*: for charting, to keep records and history, to store x-rays and notes

Factor #26: Accessible

Accessibility of a dental office is important to our well-being. A dentist too would want his office to be in a place where he can get the maximum patients. This would depend a lot on his area of specialization. It would also depend on his resources. Often, dentists start off small, and as their practice builds up, they expand. They might expand their existing office, or move to an area with better accessibility and better prospects, or open locations in other parts of the city. This, of course, would help prospective patients.

It is good to know that a competent dentist is in the area where you live, or at least in an area that is easily accessible. Traffic conditions being what they are, it would help immensely if the dental office was in a locality or area that could be accessed easily. Parking is usually another important concern. The dental office should have unrestricted parking.

In case the dental office is in a high rise building, and is not on the ground floor, there should be a lift big enough for a wheelchair, since the physically challenged and the elderly visit the dentist. Also, if the dental office has many rooms, then it would help if all of them were on the same floor. Doors should be wide enough as well so that wheelchairs can move in easily without the patients getting hurt.

Since dental treatment is viewed as part of the holistic treatment of the patient, it is good if the dental office is close to a hospital or medical clinic. Dental specialists, dental laboratories, and pharmacies, should be close by for patients' ease. A modern dentist would have his own website where his telephone number, address, and services offered, would be given. This gives regular accessibility even during non-clinical hours. He would be always accessible to his regular patients as well.

Factor #27: Warm

Going to a dentist is not always pleasant. However, you know you need to do it. Of course, regular dental visits build up trust between the dentist and the patient, so that regular dental visits are not daunting. It is those who go only when there are dental problems, who feel anxiety and fear. Modern dentists are sensitive to their patients and ensure that their dental offices are warm and welcoming. The décor of the rooms and the right climate control ensure that the office is friendly and inviting. The colors used, flowers, and piped music go towards making this a cheerful place. The reception area provides educational videos, books, magazines and wireless access to the internet. Courtesy refreshments add to the atmosphere of warmth and caring.

Comfortable dental chairs are a sign of caring. Also, in case the patient is cold because of anxiety, a soft blanket and a warm neck pillow are available so that the patient is snug and warm when being examined. If the dentist is genuine and concerned, and is a good listener, the atmosphere in the dental office will reflect that. If you are a first-time patient, you are given a tour, made to feel welcome, and made to feel that no dental problem is so big that it cannot be handled here. Regular patients are treated with courtesy and respect. This instantly creates a feeling of goodwill and warmth. Though the atmosphere is friendly and relaxed, all appointments are kept punctually.

Since modern dentistry stresses on preventive measures, how to have good oral health would be taught and talked about in an informal, relaxed manner. Comprehensive, personalized care and attention with state-of-the-art equipment is the best way to create an atmosphere that says *We Care*. The office clearly reflects the attitude of all who work there. Good, pleasant, and positive vibes go towards the creation of a warm atmosphere in a dental office.

Factor #28: Well-planned

There has to be a purpose behind every design, and a dental office is no exception. While it reflects the professional image of the dentist, it should enhance efficiency and productivity, and be a stress-free place of work. Cost is important so, how the dentist prioritizes and yet has a well-planned office will depend on his business skills. Ideally, dental offices would have:

- *Reception and welcome area*—This should have some sort of educational videos, books, magazines, flowers, and a beverage bar.
- *Consultation room*—This is the room where consultations, medical histories, and case presentations can be made and discussed.
- *Examination and Treatment rooms*–The rooms should take into consideration patients' need for privacy and comfort.
- *Laboratory*—this depends on the space available. It could also just be a small area where the dentist can keep his lab equipment.
- *Common room*—this is where the staff can have meetings, eat, and take breaks. Naturally it has to be away from the treatment rooms.

All rooms should be well-equipped according to their purpose. The equipment should be arranged so that the rooms do not seem crowded, and yet inspire confidence and a sense of well-being in the patients. This is especially so when it comes to the instruments the dentist uses. A well-planned dental office should be spotlessly clean, well-organized, and there would be no sign of clutter. Books and magazines for the patients to go through while waiting for their turn should not be outdated, torn, or tattered.

Since his office is a reflection of him, the dentist needs to make sure that his rooms wear a welcoming look. Patients, too, have a way of doing their own assessment and well-planned offices would have a favorable impact on them.

Factor # 29: Intuitive

Increasingly it is being acknowledged that environment affects our moods. Concepts like Feng-Shui and Vaastu are making people aware of their surroundings, and conscious of the arrangement of their homes and offices. These concepts emphasize the importance of space and the flow of different kinds of energies in that space. In dental offices too, it is important to follow these concepts. The first thing would be to de-clutter. Remove all equipment, material, and furniture that are not required. Clutter is enervating and exhausting.

Next is to ensure that the office is clean. The Environmental Protection Agency or the EPA has proof that the quality of indoor air is among the top 5 environmental health risks today.[44] Once all the dust, pollen or dust mites are cleaned out, the energy generated revitalizes all those who work there and those who come for dental aid. There are many new, effective cleaning agents available. Color is a source of energy. Intuitive Designers help explain the energy of different colors, so that the colors used for the walls, as well as any fabrics used, create a sense of harmony and tranquility. The kind of lighting and amount of light in the rooms also helps in the creation of a sense of peace and well-being. The energy released by the materials used in the décor stimulates the minds of all those in the dental office. The arrangement of the furniture is an important aspect of an intuitive office. Functional furniture as opposed to elaborate furniture tastefully arranged produces the right kind of energy. Each room should have only furniture and accessories specific to it. This is so that there is free flow of energy around the room. In the ultimate analysis it is Nature that heals. Therefore, if the spaces in the dental office are aligned with the healing forces of Nature, the patient benefits. The right kind of energy creates a positive environment, and helps in physical and mental healing. The wrong kind of energy brings out negativity, and increases pain and discomfort. It is wise to remember that space is dynamic. People are constantly changing. If the dentist is committed to continue learning, dynamic in his thinking, proactive and passionate about his work, it will reflect in his office, how he changes the look and feel of the office to reflect his frame of mind, and the changes he forges in himself.

Factor #30: Ahead of the Curve

Since a person going to the dentist wants the best possible treatment and dental care, it is imperative that a dentist stays ahead of the curve. For this, the dentist needs to be in touch with all that is happening in the field of dentistry as well his particular field of specialization. This enables him to anticipate issues that may come up and work out solutions, so that he is always on top of the problem. He is not satisfied with things as they are but constantly seeks to improve his dental practice by introducing new systems, new methods of treatment and by finding new solutions to old problems. He makes it his business to learn new methods of treatment, and is ready for any emergency. His knowledge helps to counter most anything negative that may arise.

A very important thing is that a dentist, passionate about his work, learns from his peers, colleagues, and literature in diagnosis and treatment without any fear to make decisions. A committed dentist will involve himself in the oral health of his local community. He is not afraid to collaborate with other medical and dental professionals. He builds long-term relationships with his patients.

It is only if he and his office stay ahead of the curve that he can become more confident and self-aware. He makes use of the latest in information technology so that he can use it for your advantage. He uses his website to provide information on all the services that he offers. He uses his knowledge to increase his efficiency and his methods of practice.

The biggest question here is that if it is such hard work and there is chance of failure: *then why bother?* The answer is that the world is constantly changing, and if you want to keep abreast of that change and even try to get ahead of that change, then it creates an environment for you to receive the best care possible. Also, changes that can be successfully incorporated into your care make for increased success rates. Thus, what is required to be ahead of the curve are:

- Passion
- Knowledge
- Determination and perseverance
- To be proactive[45]
- To collaborate with others

CHAPTER 4

FACTORS 31–40
TOP 10 ASPECTS TO LOOK FOR
IN A DENTAL TEAM

Factor #31: Cohesive

A cohesive dental team is an absolute must for a dentist. However, once he has chosen his team, it is his duty to keep the team together. Members of a dental team will include dental assistants, hygienists, laboratory staff, and office staff. It will also include, by extension, vendors, lab technicians, and other professionals who might need to be referred to. Everyone likes to be part of a team for the simple reason that everyone has a need to be part of something that is bigger than them. For a dental team to function in a cohesive manner, the dentist, who is the leader, will need to motivate and inspire them so that their morale remains high and they work together.

He needs to be open to change and sensitive to the needs of the team. A team works best when there is a positive, trusting, and facilitative atmosphere. Each member needs to feel important, to get continuing education, and to be recognized for their efforts in terms of praise. These increase the confidence and satisfaction of each team member.

Team cohesiveness is a dynamic process. Members of the team complement each other with their own specific skills. It is only when each member of the team understands his way of working as well as the style of working of his teammates that a sense of mutual respect builds up. Small adjustments are made willingly keeping the main goals and objectives in mind, and should there be set-backs, the team tackles it as one unit. Working in the team, then, becomes a pleasure.

There are certain time-tested philosophies that help here. For instance, the FISH philosophy teaches team members to care about each other, remain connected to each other, and always be aware of their commitment to a common goal.[46] The FISH philosophy also enables team members to have fun at work. It is important that everyone enjoys the work they do. This is the only way a positive work environment can be created in the dental office.

Thus, what builds a cohesive team is:

- Having a great attitude: we get to choose our attitude every morning. Your dental team should choose to have a great attitude.
- Enjoying what you do every day: Having fun
- Make our patients' day
- Being there for our patients: Physically, mentally, emotionally
- Collective responsibility and accountability
- A caring and sharing attitude towards each other, and the ability to uphold each other despite set-backs and difficulties that might be encountered
- Pride, belief and faith in the organization
- Honest and open communication between the leader and the team, as well as among the team members
- Complete trust in each other

Factor #32: Kind

A warm, welcoming, and kind dental team automatically puts patients at ease. They make each patient feel unique and important. A team that is warm and caring is so because each member of the team is valued not only for the special skills they have to offer, but because they are essentially caring human beings. They are concerned about the patient and their one and only aim is to keep the patient comfortable and happy. Warmth and caring within the dental team result in respect of one another. Each one respects the special skill that the other has, and the position of each member on the team.

The team leader, who is the dentist, has to have a clear idea of the kind of environment he wants. Knowing that a warm environment exudes a feeling of peace and happiness, he has to first of all create this warm feeling in the way he handles and deals with his team members. If the team members know that their leader is accessible, respect him for his knowledge and his work, is ready to move with the times, and is not rigid in case he has to make changes, they automatically feel a bond. The leader creates a fun atmosphere without diminishing the seriousness of purpose of a dental office. This permeates to each member of the team: a win-win.[47] A positive environment is a healing environment, and each team member needs to realize that they are responsible for each other's positive state of mind.

In case of problems or errors of judgment, the leader does not accuse the team member. Instead, he uses Ken Blanchard and Spencer Johnson's management techniques to turn a difficult situation into a learning experience.[48] Everyone makes mistakes, but everyone needs the chance to rectify these mistakes. This, in a caring environment, happens without the person losing his sense of self-esteem. In turn, the happiness translates into caring towards the team members, the leader and the patients.

It is a known fact that the vibrations created by positive thoughts, in turn makes the environment happy and warm. Instead of competing with each other, when team members collaborate with each other, patients immediately sense that they are in good, safe hands. Because the professional demands of a dentist and the dental team are high, a sharing and caring attitude translates into similar actions. Thus, without compromising on quality of dental care, all oral and dental care is administered in a kind, safe atmosphere simply because the team is warm, responsible and caring.

Factor #33: Skilled

Ideally a visit to the dentist is a fun event, where you get your teeth and gums checked. Often a dentist has a dental team where each member of the team looks after many aspects of your mouth. Since the dentist and the members of his team are dealing with such a sensitive part of the body, they all need to be highly skilled.

It may be just a routine cleaning and check-up that you visit the dental office for. In such cases the dentist or a member of the team will see if the teeth are in good condition and the gums are healthy, and assess and review your oral health. This will likely include some x-rays, a gum evaluation known as periodontal charting, an oral cancer screening, a cleaning, and exam. Sometimes it will also include use of the laser for bacterial reduction or fluoride treatment to decrease the risk of cavities.

Sometimes special care might be needed to get further diagnostic data which would involve taking a CT scan, Velscope oral cancer screening, taking impressions of the patient's teeth to make casts, and taking photos.

Besides these, there are areas of dentistry such as Cosmetic Dentistry, Restorative Dentistry, Endodontic Dentistry, Periodontal Dentistry, and Orthodontic Dentistry which require great manual dexterity and skill. Oral Maxillofacial Surgery is a highly specialized surgery. There are other areas too, which require skill, such as Pediatric dentistry, TMJ Treatment, and various kinds of dental emergencies. Skill comes from knowledge, practice, experience, and discipline. It is the expertise and skill of each team member that makes the treatment successful.

In fact, the skills required are:

- *Communication skills*—such as talking to patients to make them comfortable, asking relevant questions, and listening carefully. Communication skills are required when communicating the diagnosis to the already-frightened patient. To be able to tell the patient what is wrong, to outline the course of treatment, to help work out the financial arrangements, if required, and to give referrals if needed, require the best communication skills possible.
- *Clinical skills*—are required for examining the patient.
- *Critical thinking skills*—are required to collate and begin to interpret the data that has been obtained. This is naturally done as a team when each member puts forward their views based on their examination and tests done.
- *Treatment skills*—depending on the role of the team member, this could range from administering local anesthesia to taking radiographs or taking impressions.
- *Follow-up skills*—often after the treatment is over, the patient is forgotten. A committed dental team will never allow this to happen. A schedule is worked out for follow-up of treatment undergone as well as regular check-ups.

Factor #34: Experienced

There is no substitute for experience. A dentist and his dental team, where every member has experience, inspire confidence in the patient. It is important for you to know that the team members who you see for a check-up or treatment know their job well. While it is important to be highly qualified, and to continually keep in touch and learn, it is of vital importance to have experience. For one thing the more experience a dentist or members of the dental team have in their areas of specialization, the gentler they are with the patient. Thus, whether it is cleaning, restoring or improving the condition of the teeth, the dental hygienist, or assistant, or the dentist himself, does the work very gently. Another thing is that experience eliminates frustration. This is actually most important where the professional is concerned. It is only through experience that the dental care giver learns which procedures work best, or which kind of equipment works better than others. Even dental procedures are fine-tuned with experience. Sometimes, all kinds of things can go wrong in the simplest of dental procedures. Only those with experience can handle this kind of a situation with equanimity. The same goes in times of dental emergencies. Only those with experience will handle the issue at hand in a calm and controlled manner.

For the student of dentistry, it is important to gather as much experience as possible while in college, and after graduation. By gaining a variety of experiences in dental work and dental-work-related fields, he will get an insight into the various fields of dentistry. This helps in making an informed decision about which field of study he wants to pursue.

Experience hones all the skills that one needs or acquires to be an excellent dental care giver, whether they are communication skills, clinical skills, diagnostic skills or treatment skills. Experience teaches the significance of team work. Experience teaches a person how to be a leader as well as a team member. Getting an idea of how dental work is done in other countries also gives invaluable experience. It helps a great deal if a dental care giver gets experience in every area of dental care giving. It would be good for him to have experience in dental nursing, as an assistant, and as a hygienist. Gaining all-round experience goes towards the making of an efficient dental care giver.

Factor #35: Teamwork

There is great synergy in teamwork. The dental office may be equipped with the latest equipment and may have the latest that technology has to offer. However, for a high level of harmony and cohesion, high quality work and for the sheer joy of excellence, people must work as a team. It is a proven fact that people need to feel part of a team. There are some ground rules for a team:

- Having the same core values
- Commitment to the goals and objectives of the profession, and commitment to building a great dental practice
- Clarity of vision so that there is no room for doubt as to why each one is on the team. This is because every member of the team is working towards the same goal
- Communication so that there is an atmosphere of trust
- Pride in oneself, in one's knowledge, abilities and skills, and in the organization that one chooses to work in
- A high sense of integrity
- To be respectful with each other so that even if there are differences of opinion, the basic respect for each other helps tide over the difficult situation
- A deep sense of professionalism guides the team to keep the interests of the patient above all other considerations, and to see the patient through to the end of the treatment
- Having a sense of responsibility towards each other
- Cooperating and collaborating with each other so that everyone is in harmony with one another
- Being passionate, dynamic and proactive
- Accepting change readily, personally, and as a team
- Keeping a high morale at all times
- Remaining focused on the task in hand with an eye on the future
- Having a code of conduct which makes the team stick together in the face of any challenge or difficulty

Teamwork enables each member of the team to learn from failure. Failure is an integral part of success, and if the team takes failure in the right spirit, without blaming each other, then future patients as well as the organization benefit.

It is only a leader with a clear vision of the work he wants to do, who can put together an effective team that embodies the concept of teamwork.

Factor #36: Happy

A happy dental team views their office as a place they love being in, a place they enjoy working in, and a place where they are committed to excellence. This reflects in the environment they create in the dental office. According to *The Secret*, we attract everything into our life.[49] There are some facts that prove this. Vibrations are the law of our Universe. We are made up of molecules that are in a constant state of vibration, and we create vibration around us. We also have a mind which has the ability to consciously choose our thoughts. Thought waves are cosmic waves, which control the vibration we are in. If we choose happy and positive thoughts, the vibrations that we create around us will be positive and happy. If we are negative, we create negative vibrations around us. According to the law of attraction, our positive, happy vibrations attract similar vibrations.

Therefore, if every member of the dental team is happy and positive, those are the vibrations that are created in the dental office. Everyone who comes to the dental office will feel it. A visit to the dentist can be frightening because of the pain that the patient is suffering, or because of the sound of the dentist's drill, or the injections that he might have to take, or the fear of previous bad experiences. If the patient enters a positive, happy environment, the fear can be removed. Even if there are problems, a positive and happy environment makes it possible to cope with any difficulty. It is a known fact that even cancer has a higher rate of remission if the patient is happy and positive.[50]

Dental treatments work better on patients who are happy. The onus, then, is on a happy dental team. It is the responsibility of the leader of the dental team to ensure that each member of his team is happy. The ingredients that go into making a happy dental team are:

- Daily communication with each member of the team, as well as with the team as a whole, thus creating an open and respectful environment.
- Appreciating the strengths, knowledge, and skills of each member, and showing that appreciation.
- Inspiring confidence in each member that everyone is in this together.
- Making all criticism a constructive and learning experience.

Factor #37: Punctual

A dental office where punctuality is a value is one where time is managed in the best and most effective manner. It also ensures a high degree of effectiveness. If punctuality is a strong value with the leader, it automatically becomes so for the team as well. Punctuality starts with the time everyone reports in for work. If everyone comes in on time, the tone for the day is set. In his morning huddle, the leader would emphasize this in terms of appointments to be kept.

Time is precious for everyone, so patients should never be kept waiting. If the dentist and members of the dental team practice the discipline of punctuality, a lot of the fear and trepidation of the patient is dispelled. However, in case there is a patient with a serious dental emergency or complication, it helps if the dentist or one of the team apprises the person next in line about the problem so that he can take a decision whether to wait, or reschedule another appointment. This immediately sends out the message that the dental team cares, and is conscious of and respects the patient's time. In turn, the patient feels that he is valued. In a survey of patients, punctuality of a dentist was the most important factor in a dentist's reliability among adults.[51]

In case of a first-time patient, keeping the first appointment punctually, establishes a good patient-dentist relationship. Punctuality on the part of the dental team builds trust and respect in the patients. This is because communication is an integral part of any dental treatment. If the dentist and his team keep to the time, then there is enough time to accommodate each patient comfortably without rushing through. Every aspect needs time, from the consultation, to the diagnosis, to outlining a treatment plan and most importantly in allaying the patient's fears and building confidence. It matters little if the dentist is very skilled and knowledgeable.

If he or his team cannot be punctual, then it unfortunately goes to prove that they are not reliable, and this in the long run actually becomes self-defeating. Further, it is likely that the unhappy patient will find a new dentist.

Punctuality is important where results of clinical tests done need to be reviewed with the patient. The whole treatment plan depends on this. Punctuality decides the quality and quantity of work done.

Factor #38: Motivated

It is only a highly motivated and self-driven dentist who will be able to build a team that is also highly motivated. If he uses cutting edge technology and the latest methods of preventive, restorative, and cosmetic dentistry, it means he is not satisfied with the status quo. It proves he is committed to continued education and of using his newly-acquired knowledge and processes to better his techniques and skills. He is thus able to inspire and train his team so that they become more patient-centered and efficient in giving dental care. Automatically, stress levels are reduced, making for a hospitable work atmosphere. An inspired and motivated team is ready to take on challenges and put in extra hours of work when required, without feeling the strain.

Having a team leader who demands professional excellence and at the same time is understanding and recognizes their effort, makes each member of the team give their very best. To keep his team motivated, the leader is encouraging and supportive. Documentation of each practice or system helps in motivating the members of the team. This way, there are no knowledge gaps, and, if any changes are to be made, there is a clear idea of how to go about it.

Continued education is important for every member of the team. It is a good idea to draw up a plan for this with each member. Continued education may be in the field of specialization, or it may be attending courses in related fields. A team for whom learning is important, will be motivated enough to take on more responsibility.

Motivation does require some form of assessment as well. The dentist needs to address issues head-on. When there are issues, the matter on hand can be discussed in a light-hearted atmosphere, concentrating on the strengths and contributions made. The areas that need improvement can be talked about in a pleasant manner, without damaging the self-esteem of the person concerned. For the team member, this is of great importance because this is what will decide how he keeps the team motivated. Also, during the discussion, the dentist could ask for the team member's feedback. This kind of give and take, gives every member of the team a feeling of ownership, and consequently draws out the best in them.

Factor #39: Courteous

There is no substitute for courtesy. Politeness and consideration of others are the hallmarks of a courteous person. It is a pleasure to interact with a dental team where every member is courteous. Basic manners are absolutely essential.

A dental team handles all areas of a dental office, starting with the office administrator. An office administrator, who has good manners while speaking to those who come to the dental office, whether they are prospective patients, regular patients, other medical professionals, medical representatives, vendors, or visitors, conveys the courtesy that can be expected of every member of the dental team. The same degree of courtesy is extended when speaking on the telephone.

The leader of the dental team treats everyone, regardless of age, or seniority, with the same politeness and courtesy. He retains his respect even in an emergency thus being calm and strong when it is needed the most.

Team members must treat each other with courtesy as well. Since each member is qualified to handle a different aspect, working together means respecting each other regardless of personality differences. At all times it is to be remembered that the patient is all-important. Thus, courteous behavior would include being transparent, keeping personal and professional issues totally separate, and maintaining the highest ethical standards.

Should there be differences of opinion, they can be handled without bringing in personalities and without drama. While dealing with patients, remember that all kinds of people visit a dentist, from teenagers to adults to the elderly.

The dentist attends to the physically and mentally challenged as well. Therefore, it is imperative that all patients are treated with the best possible manners and the highest consideration. Patients should always be addressed by name, and spoken to with respect and courtesy.

Factor #40: Conscientious

Conscientiousness is probably the greatest defining characteristic of a dental team. This is what decides how good the dentist and his team are. Every member of a conscientious dental team shows great care in carrying out their own individual as well as collective responsibilities. They go about their work in a purposeful manner.

They are conscious of the fact that they are dealing with a very sensitive part of the body, and also with the inherent fear that a visit to the dentist involves. So, they ensure that the patient is physically and psychologically comfortable before any kind of consultation or clinical examination takes place.

They ask detailed questions and pay attention to every little detail so that nothing is missed out. During the clinical examination, all procedures are scrupulously followed and whatever has to be done, is done meticulously and thoroughly. If every member of the team does his work sincerely and conscientiously, then the dentist will not need to double check. He can rely on the work of his team and arrive at the right diagnosis. There are certain qualities that a conscientious team has:

- Self-discipline
- Ability to work hard and selflessly
- Responsibility
- Reliability
- Organizational abilities
- Orderly procedures and practices
- The ability to be able to think carefully and deliberately without hurrying
- Persistent and persevering
- The ability to collate data
- The ability to question and also to rectify errors
- Consciousness of the goals to be achieved
- The need to achieve

While outlining the treatment plan, whether it is a routine cleaning of the teeth and removal of plaque, or performing some aspect of cosmetic dentistry, the same thoroughness and care is given to both. There is no hurry in the work that is being done. There is no hurry in discussing prevention of dental problems and how to go about oral hygiene.

For a conscientious dental care giver, no detail is too small or too insignificant. He will work as thoroughly without supervision as he would under the watchful eyes of the dentist. A conscientious member of the team would feel responsibility for the work of the other members. If and when mistakes are made, the conscientious team member will not keep quiet, but will, in an atmosphere of collaboration and responsibility, work towards rectifying that mistake. This is what will set and ensure high standards.

CHAPTER 5

FACTORS 41–50
TOP 10 PIECES OF TECHNOLOGY
YOUR DENTIST SHOULD HAVE

Factor #41: i-CAT—Cone Beam CT Scanner

More and more, dental practitioners use the Cone Beam computed tomography (CBCT) to obtain 3-dimensional digital images of the region of interest within the oral cavity.[52]

CBCT is a device that produces three-dimensional imaging within the oral and maxillofacial complex. While the patient sits in an open environment, the scanner rotates 360 degrees around the patient's head in 9 to 25 seconds. It is important to note that the radiation absorbed by the patient is approximately 50 times less than a traditional medical CT scan.[53] Finally, the reconstructed data is imported into software that provides high-definition, 3-dimensional images of all oral structures.

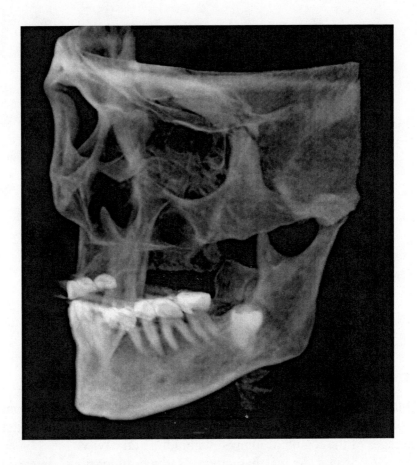

CBCT is an important diagnostic tool in dentistry because lesions can be detected in their early stages. Furthermore, the additional information gained from three-dimensional images compared to two-dimensional images allows the practitioner to perform a more accurate treatment plan and thus better treatment for the patient.

In more detail, the radiation acquired from the patient (raw data) is reconstructed into a folder of digital images that are called DICOM images. The DICOM images are imported into viewing software where the practitioner may manipulate the data to view hard tissue structures. The data does not exhibit distortion, magnification, or overlap of anatomy, commonly seen in two-dimensional images. How is this better for the patient? The acquisition process is fast, painless, and the patient absorbs less radiation compared to medical CT. Within minutes, the practitioner can communicate and discuss regions of interest with the patient that they can understand. This procedure is less expensive than the cost of a medical CT exam.

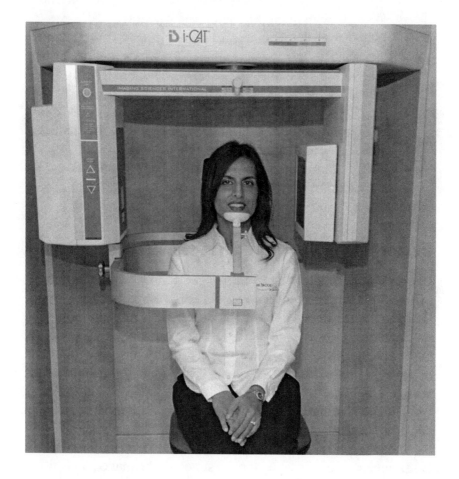

Limiting factors for more practitioners to obtain this technology is the cost of equipment and additional training which is required to use the data correctly.

How does CBCT help the dentist?

- The practitioner obtains immediate 3-dimensional information.
- CBCT information is used by Dental Implantologists, Orthodontists, Periodontists, Oral Surgeons, TMJ Specialists, ENT specialists, Allergists and General Dentists with great efficiency and success.
- Because of the 3-dimensional re-construction, the dentist can analyze the bone structure, tooth orientation, critical bone-tooth relationship, Temporomandibular joint (TMJ) disorders, impacted teeth, root fractures, problems with the 3rd molars or wisdom teeth, periodontal conditions, congenital defects, oral-nasal passage, and the paranasal sinuses, or any other disorder that may be present.[54]
- CBCT data can limit procedures for patients. Also, the time required for surgery can be reduced.

Factor #42: Velscope—Oral Cancer Screening

Cancer statistics show that oral cancer claims twice the number of lives as compared to cervical cancer. This is because, by the time the patient decides to see the dentist, it is already too late to do anything. However, if it is detected early enough, there are 80% to 90% chances of recovery. This is why it is very important to have an oral cancer examination done annually, as part of the normal oral hygiene examination.

The most common causes for oral cancer are tobacco, alcohol and an indiscriminate sexual life.[55]

Cancer is the abnormal growth of cells. This occurs in the basal membrane and it is not visible to the naked eye till it reaches the top, by which time it is too late to do anything.

The Velscope system is a cancer screening device. It is *not* a diagnostic device. Velscope uses fluorescence visualization technology to detect pre-cancer and cancer lesions, as well as any abnormal tissue that might be in the mouth. The procedure takes only 2 to 3 minutes, is non-invasive, and is completely painless. The patient suffers no discomfort at all. Also, this procedure is neither messy nor difficult for the dentist or hygienist to administer.

Before doing the Velscope examination, the dentist does the conventional White Light examination. In this, the dentist looks for lesions, and palpates the neck and face to check for lumps. In order to remove all doubt, the dentist does the Velscope examination since pre-cancer abnormal tissues are not visible to the naked eye. The Velscope hand piece, which emits a safe blue light, is used to examine the oral cavity. This light causes the tissue from the surface of the epithelium through the basal membrane and down to the stroma that is beneath it, to become fluorescent. If there is no problem, the whole area shows a green fluorescence pattern.[56] However, if there is any abnormality, there is a change in the fluorescence and the abnormal tissue or lesion stands out as dark and irregular areas. If the dentist has a doubt about the abnormality, he recommends that a surgical biopsy be done. The biopsy is evaluated by an oral pathologist, and then a treatment plan is worked out.

In case of surgery, the Velscope is used by the surgeon to identify the diseased tissue around the lesion. This helps him to work out and detail his surgical plan.

The Velscope is an important tool to detect and prevent oral cancer. Make sure your dentist is performing an oral cancer screening, and ask about Velscope.

Factor #43: Diode Laser

The dental laser is the latest in technological devices that helps in soft tissue management.[57] Laser energy is magnified light. The energy source is attached by an optical fiber to a hand piece. The laser energy is produced in pulses and as these pulses leave the optical fiber, they make a small ticking sound. The diode laser has the following clinical applications:

Laser Bacterial Reduction (LBR)—Disinfect the gums and stop the bleeding when your pockets are 1–3mm.

Deep Laser Gum Therapy (DLGT)—When pockets are 4mm or greater, DLGT removes infected gum tissue without the need for invasive periodontal surgery. It also enables you to get regrowth of bone and attachment around teeth.

Gumline cavities—Cavities at the gumline can be restored without any contamination by infected gums. Diseased gum tissue can be removed and bleeding will not occur.

Exposure of unerupted teeth—sometimes teeth do not emerge as they should because they are covered by gum tissue. Laser technology helps uncover the teeth without damage to the surrounding areas.

Frenectomy—the extra folds of tissue that restrain the movement of the lips, cheeks, or tongue are released without bleeding and without sutures.

Gingival troughing for crown impression—get clear impressions of the exposed crown margin without any bleeding

Gingivectomy—it is possible to easily get impressions for crown and bridge procedures. Redundant tissue is removed without loss of blood and painlessly.

Implant recovery—extra gingival tissue around the implant can be removed easily and safely.

Operculectomy—this is the removal of redundant soft tissue that is behind the molars.

Post Orthodontics—problems arising after de-banding are successfully treated with the ZAP laser

Soft tissue crown lengthening—also helps in crown and bridge procedures.

Desensitization—Dental sensitivity is most often caused by gum recession (when the gums migrate away from the enamel of a tooth). Lasers have been proven to be effective at decreasing or even eliminating dental sensitivity. A simple procedure requiring no anesthetic directs the laser energy into the sensitive areas and dramatically decreases sensitivity.[58]

Sulcular debridement—is the removal of any diseased soft tissue that might be found in the periodontal pockets around teeth.

Lasers have been specifically designed to treat periodontal disease. Periodontitis destroys the tissue as well as the bone that supports the teeth. The trouble with disease of the gums is that it is painless and so the person is not aware that something may be wrong.

Dental lasers help treat a number of diseases and prevent the need for a scalpel in many cases. Ask your dentist about laser dentistry.

Factor #44: Digital Radiography

It was in the 1980s that the first intra-oral sensors that could be used in dentistry were developed. Through the years cost-effective intra-oral and extra-oral digital technology has been developed, and this along with advancements in computerization, has made digital imaging superior to the conventional film imaging. In digital radiography, the image is created using pixels which are tiny light-sensitive elements. They are arranged in grids and rows on the sensor in shades of grey. The signals produced by the sensor are analog signals. These signals are communicated to the computer in analog form. This analog data is converted to digital data, and after being processed by the computer, is displayed as a visible image on the computer screen. These images are in shades of grey. There are 256 shades of grey. 0 is black and 256 is white, and the numbers in the middle represent all the shades of grey. The number of grey levels and the size of the pixels determine the resolution of the image formed. Direct digital imaging thus produces a dynamic image which can be enhanced, stored, retrieved, and transmitted. Digital sensors require much lower radiation exposure than the conventional method.

With digital radiography, images are available instantly on the dentist's computer. He can show these to his patient and discuss the diagnosis and treatment plan with the patient. He can also send these images to his colleagues in case he wants to further discuss the diagnosis. The patient can have all the x-rays on a compact disk, as well, should he want to go to another dentist for a second opinion. Being able to view the images instantly is of especial use in endodontic therapy, implant surgery, evaluation of the crown fit, placement of posts in teeth that have been treated endodontically (i.e. with root canals), evaluation of newly placed restorations, detection of objects that are radiopaque (white or with high attenuation), and for the oral education of the patient.

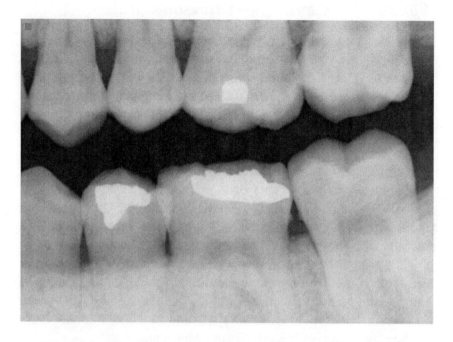

The fact that there is less radiation in digital radiography, allows for multiple images in a short exposure time with almost no risk to the patient. For example, a full set of x-rays is equivalent to the radiation we normally receive from the world we live in over the course of 2 days.[59] That is more than 10 times less radiation when compared to traditional dental x-rays.

This is of great use in placement of implants and during difficult endodontic therapy. The software used in digital radiography is easy to learn and master. An important consideration is that digital radiography is easy, clean and fast.

Make sure your dentist is using digital radiography in order to get the most diagnostic imaging with the least radiation exposure.

Factor #45: Cavitron—Ultrasonic scaler

Daily cleaning of the teeth, good dental habits, oral hygiene and regular visits to the dentist are of paramount importance. The first sign of neglect is the build-up of plaque just below the gum line or the sulcus. If the plaque is not removed regularly, it hardens to form tartar. If still left unattended, the tartar causes the gums to separate from the teeth, forming pockets. The pockets fill up with more plaque. This eventually leads to periodontitis or gum disease. If the situation becomes more severe, then the dentist will have to perform surgery on the gums.

All this can be avoided by regularly removing the plaque and tartar from the teeth particularly along the root surface. The method of removing plaque is called scaling. Scaling used to be done by hand-held instruments. New instruments though have diminished the need for these hand-held scalers. Now, scaling is done with a device called a Cavitron, which is a state-of-the-art ultrasonic scaler.[60] It looks like a wand and has a small blunt scaling tip through which the sound waves pass.

Cavitron uses these high frequency vibrations of sound to literally blow away the tartar from above as well as from below the gum line.[61] A high speed jet of water is sprayed along with the vibrations thus cleaning the teeth as well. The action of the Cavitron is so gentle, and yet so effective, that deep cleaning is possible without an anesthetic. Since it cleans the teeth quickly, the patient is quite comfortable with this method of tartar removal. The stress factor, too, is sufficiently reduced, both for the dentist and the patient. An important point to remember would be not to use the Cavitron in case the patient has a pacemaker. The vibrations of the Cavitron would interfere with the pacemaker.

The Cavitron Jet Plus Scaler is an ultrasonic scaler and air-polishing unit. Advanced Cavitron scalers have a 360 degree wireless footswitch, illuminated diagnostic display, rinse setting to wash away the debris and an automated purge which ensures that there is fresh lavage for each patient. The power can be adjusted for comfortable sub-gingival scaling.

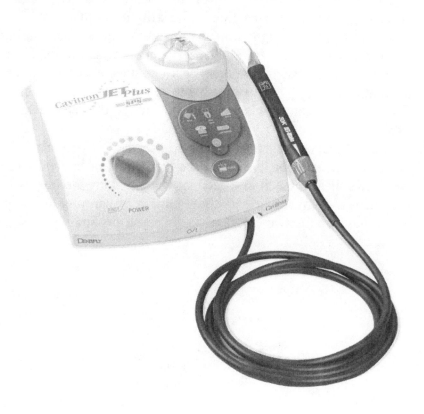

Make sure your dentist and hygienist are using a Cavitron.

Factor #46: Dental Loupes

Dental loupes are devices used for magnification. Dentistry is a difficult science, since it is challenging to clearly evaluate the oral cavity. Using dental loupes your dentist and hygienist can inspect the inside of the mouth clearly, and focus directly on the specific areas. Dentists need to use both hands, so dental loupes are fitted like glasses. Flip-type loupes look like cylinders and are fitted in front of each lens of the dentist's or hygienist's spectacles. Loupes can be fitted with a light, so that there is a continuous source of light aimed at the area of importance during all dental procedures. Multiple lenses help especially in the fields of implant dentistry, oral surgery, periodontics, endodontics and restorative dentistry.[62] It is possible to customize the dental loupe for better and sharper clarity, and to suit the dentist's field of expertise.

Dental loupes are custom made so that they fit properly. The power of the eyes, or lenses in case the dentist or hygienist use spectacles, is also taken into account when making the loupes so that there is optimal clarity without hurting or harming the eyes. Since these are lightweight, they do not cause any strain.

Dental loupes are of great use when the dentist is examining a decayed tooth. While drilling, since the tooth is magnified, it is easy to see for how significantly the tooth is decayed, thus helping in precision drilling. In fact, all dental procedures are made more accurate if done using dental loupes. New models feature an LED light system.[63] This illuminates the oral cavity with a white, bright light. There are no shadows in this kind of light, and no strain on the eyes at all. New technology makes it possible to have cameras and video recorders fixed to the loupe. Another innovation is Laser Loupes. These are through-the-lens loupes with built-in protective filters in the carrier lenses and the telescope, for those who use laser technology in their dental work.

Factor #47: Piezo Sonic

The Piezo Sonic is an amazing innovation in dentistry today. It has revolutionized the way surgery happens. The Piezo technique is a minimally invasive technique, and causes the least possible trauma to the soft tissues that surround the bone, nerves, blood vessels and mucosa. It lessens the damage to the osteocytes (bone cells). It also allows for the survival of the bone cells.

Piezoelectric surgery uses ultrasonic vibrations to complete a procedure. This kind of surgery has been used successfully by dentists and oral and maxillofacial surgeons for gum and bone surgeries.

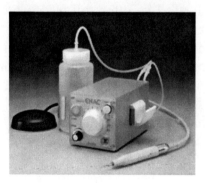

The Piezo sonic generator uses the theory that certain crystals produce a voltage when a mechanical pressure such as sound vibrations are applied on them.[64] The reverse is also possible (i.e. an oscillation or vibration takes place when voltage is applied). No heat is produced during this procedure. The Piezo tool has a hand piece into which the specific tip required for the dental procedure is fitted, and tightened. The Piezo tool can be set in such a way that it cuts only the tissues that you want to cut. For instance, large tips are used for removal of heavy calculus, and cruet tips for the removal of fine calculus. Thin tips with high power can be used on hard calculus that has accumulated in deep periodontal pockets. Thus it is used for scaling. It provides information on the condition of the gingival tissue, and the quality and extent of the attachment of the epithelial tissue with the tooth.

Though it scales the teeth effectively, it does not scratch the surface of the teeth. In fact, ultrasound technology can be used to measure the depth of the pocket as well. This helps in dental implantology. Piezo sonic technology is used in endodontics (root canal therapy) and wherever high ultrasound energy is required, such as in the condensation of inlays and removal of posts and crowns. One of the most important features of this procedure is that it is completely painless.

Importantly, the computer software sorts out the echoes and gives the accurate picture of the problem area. This technology can also be used effectively for full-mouth probing. It is quick and easy and causes no distress at all to the patient. Modern dentistry is based on the principles of maximum preservation and minimum restoration and ultrasonic technology helps in the realization of this. Piezosurgery was invented by Dr. Tomaso Vercellotti MD, DDS.[65]

Piezo sonic has made dentistry less painful and less invasive. Ask your dentist if he has Piezo sonic equipment.

Factor #48: GoldenMisch Physics Forceps

Tooth extraction can be one of the most traumatic of dental procedures. However, by using GoldenMisch Physics Forceps, there is almost no trauma at all. With these forceps teeth can be extracted non-surgically in minutes. Importantly, there is no stress at all. The principle used in these forceps is that of first-class levers, where the fulcrum is located between the input and the output. All teeth can be extracted with these forceps except for lower wisdom teeth, no matter what condition they are in.

While the upper teeth do not need to be sectioned (separated), the lower molars might need to be sectioned.

The benefits of using a GoldenMisch Physics Forceps are:

- It eliminates any root tip fractures
- Preserves the buccal bone, which is extremely important when placing a dental implant into that site[66]
- Supports implants since the alveolar bridge is preserved
- Eliminates the need for cutting the gums open
- There is hardly any post-operative discomfort
- The point of extraction heals quickly

Factor #49: Flat Screen Educational Monitors

In order to effectively communicate with you and to visually show you what is going on in your mouth, computers using flat, Liquid Crystal Display (LCD) screens should be used in your dental office. From photos to x-rays to educational videos, you should be able to clearly see what your doctor has found, why it should be treated, and what that treatment involves.

The LCD computer screen uses Thin Film Transistor or TFT technology. This produces a digital image of very high quality. The smallest single component of a digital image is a pixel. Each pixel is made up of 3 layers of red, blue and green transistors. A 15 inch screen has 1024 x 768 pixels, and a 17 inch screen has 1280 x 1024 pixels.

The advantages of using flat screen LCD monitor are:

- They have a larger viewing area, and it is easy to discuss the diagnosis with you. You get to see exactly the condition of your teeth and gums.

- These monitors use very little desk space or wall space and leave more functional work space area.
- They do not flicker.
- They are brighter and have better colors. There is a polarizing filter which sharpens the contrast, so no light gets diffused.
- The backlight is even throughout.
- Less glare and therefore less strain on the eyes.
- Uses 60% less energy than a CRT (Cathode Ray Tube), and is more environmentally friendly.
- The adjustable monitor allows for more flexibility and it is easy to discuss the diagnosis with you.
- The monitor can also be adjusted according to the height of each user, so that work posture is improved and there is no feeling of physical strain or discomfort.
- Further, the flat panel monitor can be mounted from the ceiling or on the wall.

Ideally there should be two monitors in each dental suite. One is placed in front of you so that you can see digital x-rays, intra-oral and extra-oral photos, and videos.[67] The other monitor would be placed behind your head from where the dentist and assistant can see it clearly. Naturally, sensitive material and other patients' information is not shown to you. Flat panel monitors are used while educating you about your procedure, your diagnosis, and on prevention of disease.

Factor #50: Light Curing/Whitening Device

A smile is the best way to brighten someone's day. It is also a confidence booster to know that you have a beautiful smile, and beautiful teeth. Dental enamel is discolored by multiple factors including: age, smoking, tea, coffee, and cola soft drinks.[68] Other factors, such as disease and certain kinds of medication, may stain the teeth. There are, however, people who have naturally brighter enamel. For whitening your teeth, it is best to get advice from your dentist.

In order to whiten your teeth in a dental office, your dentist must have a whitening device. Light-based whitening technology uses a special light along with a whitening gel. The teeth are first cleaned, and the plaque removed. The gel is applied to the teeth, and a special light is shone on them. The strength of the gel is determined by the dentist according to the amount of discoloration there is. The light activates the crystals in the gel, which absorb the light energy, and penetrate the enamel of the teeth.[69] The length of time required depends on how badly discolored the teeth are.

On an average it takes about an hour to get the teeth looking white. The teeth can be whitened by at least 10 shades. This method brings about an immediate result, is long-lasting, and because it is performed by the dentist, the patient is assured of safety. There is no discomfort in this procedure.

The bleaching agents used are carbamide peroxide or hydrogen peroxide. Whitening gel also contains water and glycerin so that the teeth remain hydrated.

Those with sensitive teeth will need more than one sitting, since they might find the procedure painful. In case there is gum disease, this has to be treated before going in for the whitening procedure.

After-whitening precautions have to be taken as well. For 24 hours after the whitening procedure, the patient must not have any beverage that is either too hot or too cold. You will need to cut down on tea, coffee and smoking.

Light-based whitening procedures are not the same as laser procedures. The light-based whitening procedures are based on LED technology. The light does not act directly on the teeth. It activates the gel, which then brings about the whitening.

Whitening procedures do not have an effect on dentures, veneers, dental fillings, crowns and caps.

CHAPTER 6

FACTORS 51–60
TOP 10 AMENITIES YOUR
DENTIST SHOULD OFFER

Factor #51: Coffee/Tea/Juice/Bottled Water

Do you like going to the dentist? Do you feel warm and fuzzy when you walk in the door? For a first time patient, especially, going to the dentist can be quite frightening and daunting. One of the things that puts people at ease immediately is the offer of hospitality. It is a known fact that any kind of fear or apprehension causes the mouth to dry up, and the heart to start beating faster.[70] Sipping a glass of water, coffee, tea, or juice has a calming effect. You can actually feel the stress oozing out of your system.

Any pain is aggravated if the person is in a stressed condition. Likewise, even the simplest dental procedures can seem long and painful. It is better to calm down before going in to see the dentist. Modern dentists must pay a great deal of attention to the welcome and reception area of their offices. They make sure that a soothing atmosphere is created and hospitality is offered. Welcoming a visitor and offering hospitality, is actually giving him or her the respect and courtesy due to any human being. It goes towards making you feel that you are valued. It also goes to show that the dental team is aware that coming to the dentist is not easy, and that they are doing their best to make you physically and psychologically comfortable.

For the visitor too, this is a way of knowing what kind of dental office he has come to, and what kind of service he is going to get. It is only a dentist who values his profession who will make his dental office a welcoming place. It is also a proven fact that there cannot be isolated pockets of efficiency. The dental office is taken as a whole. If there is respect being shown in one area, it will be shown in other areas too. The same respect and courtesy is bound to be shown during all the dental procedures that need to be done. Hospitality creates an atmosphere that is at once open and warm. The patient feels that no matter what he will be looked after and cared for here. A dentist's reputation spreads by word of mouth.

Do you want to go to a dentist who cares about your well-being? Most would rather go to the dentist who is warm and welcoming rather than a cold and unfriendly one.

Factor #52: Ice Cream

I scream for ice cream! Ice cream is a feel-good food. Having it in the dental office is a definite way of ensuring that people who come for dental treatment are going to be happy when it is done. Most adults cannot resist ice cream.

Dental materials don't taste great, and a great way to remove the dental taste is to have ice cream after the dental work is over. Ice cream is also a good way to reduce swelling after a long procedure. Since cold numbs the mouth and reduces blood flow to the area where cold is applied, the pain is lessened.[71] The cold makes the blood vessels contract, thus minimizing the bleeding. In fact, it is a good idea to have ice cream for 24 hours after dental extractions so that pain is minimized, and the gums can heal themselves.

Besides, of course, the fact that ice cream is an all-time favorite with everybody, it supplies Calcium to the body, and Calcium is necessary to build strong bones and teeth. In fact, 99% of the Calcium in our bodies is in our bones and teeth.[72] The Calcium in the teeth generally stays there, but the jaw bone gives up its Calcium (like the other bones) in case it is required in some other part of the body.

If the adult or child concerned is not fond of drinking milk, then one sure way to get Calcium into the body is by having ice cream. Calcium-fortified ice cream is available in the market for those who need high doses of Calcium. Calcium is, today, considered the wonder nutrient. It builds bones, controls blood pressure, reduces the risk of colon cancer, reduces hypertension, and may even prevent the formation of kidney stones.

If your dentist offers you ice cream after a dental procedure, say YES!

Factor #53: Neck Pillows/Eye Pillows

One of the major features of a modern dental office is the comfort factor. Dentists and their dental team are aware that unless their patients are physically comfortable, the whole procedure, starting with the consultation, can be a tremendous strain on both the patient and the dental care giver. Why we need neck pillows in the dental surgery is because, sometimes, dental procedures take time. They can also be quite stressful. Lying with the head on a comfortable pillow eases the tension of the neck and shoulders, and prevents the head and back from paining. There is a wide variety of neck pillows to choose from. Inflatable pillows, like the water pillows can be therapeutic. If the patient is comfortable with a hard pillow, more air or water needs to be filled in. Likewise, if it is a softer pillow that is required, then the amount of air or water that is filled can be regulated to ensure maximum comfort.

Pillows can also be warmed for a few seconds in a microwave. Warm pillows are extremely comfortable and soothing. A wrap pillow is a U-shaped pillow, which goes around the neck. They are soft, lightweight and have a pocket which, when the pillow is around the neck, is near the nose. Healing herbs can be placed in this pocket.

The use of aromatherapy in healing and reducing anxiety is being increasingly recognized. Lavender has been proven to reduce the anxiety levels in dental patients.[73] The neck muscles are the most strained when undergoing dental procedures, so a warm neck pillow with an aromatic herb like lavender is just what you want. Certain kinds of foam make comfortable pillows as well.

Eye pillows offer a great degree of comfort too. These are lightweight, and have a gentle soothing effect on the eyes. They can be in the shape of a mask, or may be rectangular. They completely block out the light as well as the sight of the dental instruments and needles. Eye pillows are filled with flax seed, lavender or chamomile flowers, eucalyptus leaves or rose petals. The cloth used to make these pillows is usually very soft, e.g., silk. These herbs block out all the smells usually associated with the dental office. They have the property of relieving anxiety and stress and actually relaxing the muscles of the head and neck.[74]

Factor #54: Hot Towels

Do you like going to a nice Japanese restaurant? The lovely Japanese custom of *Oshibori*, or offering hot towels to their guests, is used very effectively in dental offices. There is something very soothing and refreshing about a hot towel. It can be used for the face or just to warm and clean the hands. Either way, it is instantly relaxing. All that is needed are small white face cloths, water, and heat. The towels that are used should be

100% pure cotton with a luxurious texture. Rolled neatly, it is a great way to clean up after a dental procedure and extremely soothing as well. Most dental offices will have these in a towel warmer or they will heat them in a microwave. After the dental visit is over, warm towels are used. These allow every patient to leave with a fresh, clean feeling. Studies have also shown that anesthetic lasts longer when heat and moisture are added to a site where anesthetic was given.[75] You know your dentist wants you to feel great if he offers you a hot towel after a dental procedure.

Factor #55: Fresh Flowers

Which woman doesn't love flowers? What is the number one Valentine's Day gift? Fresh flowers bring a smile to most everyone. No one can deny the feeling of happiness and vitality that fills the heart on seeing a bouquet of fresh flowers. Fresh flowers are extremely beneficial in a dental office. No matter how apprehensive one is about a visit to the dentist, the sight of fresh flowers as one enters the office instantly removes that anxiety. The colors, fragrance, and arrangement of flowers bring instant joy to all the viewers. One groundbreaking study actually proved that flowers helped improve healing for surgery patients.[76]

Fresh flowers can be placed in smaller arrangements on the front desk and near the juice & coffee bar. A small spray here or a sprig there adds color and fragrance. By giving out positive energy and oxygen, these flowers remove negative energy and carbon dioxide from the rooms where they are placed.

Flowers make us smile, and smiles, as we know, are contagious. Smiles also reduce stress and cause endorphins to be released in our bodies. This immediately makes us feel better, and brighter. Tests have shown that the presence of flowers decreases physical discomfort.[77]

Tests have proven how effective flowers are in creating a positive and cheerful frame of mind.[78] In the first test, it was seen that children and women always smiled when they saw flowers, and the upbeat mood remained for almost 3 days afterwards. In a second study, both men and women were given a flower, and they all responded with a genuine smile and showed the need to engage in a conversation. In the third study, men and women over 55 years of age, when given flowers, showed an instant uplifting of their spirits, felt appreciated, and behaved more sociably[78]. All of these categories of people visit the dentist. These studies prove that you want fresh flowers at your dentist's office to help with your recovery, to decrease any discomfort, and to help keep you feeling upbeat!

Factor #56: Updated Books and Magazines

The joy that one gets in picking up a great book or magazine and going through it is irreplaceable. The very act is therapeutic. The feel of a book or magazine, the weight of it in your hands, the colors and pictures, help boost your spirit. The welcome area in a dental office is very important. The arrangement of the chairs, the décor, fresh flowers, coffee & juice bar, books, and magazines all reflect the sensitivity of the dentist.

He has to make allowances for the cultural differences of his patients, so that they are comfortable both physically and psychologically. Thus, the arrangement of the chairs will be such that people who want to sit together can do so, and people who want to be by themselves will be able to do so as well. Finding a comfortable place to sit, and reading or thumbing through a book or magazine distracts you from what awaits them in the dentist's chair. It is absolutely imperative that all books and magazines are recent or current issues. Out-dated magazines send out a message of indifference, which is the last thing a patient wants to feel.

Books and magazines make for absorbing reading because there are articles that everyone can relate to. Sometimes, the dentist might have a specific medical or dental journal displayed so that interested patients can read and make themselves knowledgeable about the latest developments in these fields. Having the latest books and magazines in the dental office keeps you occupied while you are waiting for your turn.

No matter how punctual the dentist likes his appointments to move, there are times when an emergency or complication occurs. In such cases, you will not mind waiting if there is a hospitable, comfortable place to relax, with books and magazines that you can flip through. Make sure your dentist keeps current books and magazines.

Factor #57: Educational In-Office Videos

Modern dentists use educational videos to explain dental procedures. These are far more effective than mere talking and explaining. Words could lead to confusion, repetition and the uncertainty of whether the patient has understood what was being said. This is especially true when the dentist starts talking like his professors talked during dental school, using HUGE words like: periodontal ligament, epithelial attachment, gingival hypertrophy, periapical lesion, and supernumerary dentition (these are all real dental terms: Google them).

There is absolutely no confusion in a simple, visual explanation. There is software available that shows dental procedures in animated forms. Three excellent dental educational software programs available are Caesy, NobelVision, and Implant Vision. When the relevant video is shown to you if you need some type of dental treatment, you are able to see and understand what is going to happen.

Videos are available on all aspects of dentistry such as General Dentistry, Implant Dentistry, Oral Surgery, Periodontics, Orthodontics, Cosmetic Dentistry and Endodontics.[79] Of course these 3D animated educational presentations are not a substitute for the actual clinical examination and consultation. These are used to help you understand what is going on.

It is a known fact that if you can actually see the problem and the treatment, you will have less anxiety and apprehension.[80] Patients would feel that they are actively participating in the whole procedure along with the dentist. Educational videos on the anatomy of the tooth, the different kinds of teeth and their arrangement on the jaw bone, and other related issues, help children and adults learn what is in their mouth. Videos on dental hygiene are especially useful for everyone, as they teach good oral care in easy-to-understand, step by step, and enjoyable animated form.

Besides the educative and informative videos, there are cool videos available that are appropriate for the welcome area in the dental clinic. These are carefully chosen for their clean, healthy fun so that people can enjoy them while waiting to go in and see the dentist. If these entertaining videos have something to do with teeth, or dentistry, it helps take away the anxiety from the thought of being the next in line. To this end, there are monitors fitted on the walls in the welcome area. If the visit to the dentist is made more attractive and pleasant, you are not going to hesitate the next time you have to go to your dentist.

Factor #58: Music with Headphones

Helping you overcome anxiety is a primary goal of your dentist. In fact, dental diagnosis and dental procedures need to be customized for patients showing signs of dental anxiety. One extremely effective method of reducing and even eliminating stress is by having relaxing music in the clinic. Soft music instantly soothes the mind and body. Add to this the availability of headphones, and the patient can even listen to his own music when undergoing a dental procedure. The advent of Pandora allows you to choose the music genre that you enjoy.

The sound of the drill is one of the most feared elements of dentistry. If you have headphones on and are listening to either the music of your choice, or the relaxing music offered by the dentist, you will barely be able to hear the sound of the drill. Headphones with music work just as effectively for those undergoing oral surgery.

What happens when the patient listens to music? Music affects metabolism and this causes a change in behavior.[81] It touches the nervous system and influences the brain waves. The overall influence of these changes results in a calm and positive state of mind.[82] Research has proved that classical music is the most relaxing.[83] Patients who listened to classical music, either through headphones or absorbed it from the atmosphere around them, showed a lower level of tension and were more open to treatment. They showed better signs of adaptability and adjustability in the clinic. Dentists who use music therapy for their patients will find working with anxious patients less stressful.[84]

iPods are now more popular than the traditional CD players, though there are patients who prefer their own CD player. Music reduces the amount of sedation needed and is known to actually lower the blood pressure of elderly patients.[85] Thus we see that New Age Music, Classical Orchestral Music, and Piano Music ensure that the patients remain calm and relaxed. Ask your dentist if he offers music with headphones or Pandora during treatment.

Factor #59: Blankets/Snuggies

A feeling of anxiety will make you feel physically cold. This is especially true in the dental chair. No anesthetic is as effective if you are stressed, anxious or tense. In order for you to get your treatment done comfortably, you need to feel physically comfortable. Anxious adults respond positively to the hugging effect created by a blanket or a snuggie. A snuggie is a blanket with sleeves.

When you are wrapped in a blanket or a snuggie, the warmth and pressure causes a release of serotonin and endorphins.[86]

Serotonin calms and endorphins stimulate the brain producing positive emotions and happy feelings. That means you are more calm and content. Once you are relaxed, it is easy for the dentist to do the clinical examination and continue with whatever has to be done. Snuggies are available in sizes that fit small and large adults. While a blanket covers the patient completely, it can slide off. A snuggie, on the other hand, has sleeves, so the arms are free and you can adjust your headphones, or pick up your cell phone while waiting for the anesthetic to kick in, or use your laptop, all the while remaining snug and warm.

In a dental office, the air conditioning is adjusted to an average temperature which is comfortable for everyone. However, if you have low blood pressure or are anemic, you are going to feel cold. Since the temperature cannot be changed, giving you a blanket or snuggie makes sure you are warm and comfortable. Ask your dentist for a blanket or snuggie if you are anxious or cold.

Factor #60: Aromatherapy

Do you like the smell of a dental office? Does it trigger any past experiences for you? Modern medicine believes in the holistic treatment of patients. Aromatherapy is being used more and more to help in this method of treatment. Yet, Aromatherapy is not new. It goes back 3,500 years to ancient Egypt, where aromatic herbs and woods were used in making medicine.[87] Aromatic medicine was believed to have magical powers. Aromatherapy is based on the premise that fragrances affect the psychological state of a person.

We know that how we feel psychologically affects our physical health. Since visits to the dentist create anxiety and fear, it is important that this stress be removed when the patient visits the dentist. Aromatherapy uses aromatic essential oils and volatile plant oils. When they vaporize, the tiny aromatic molecules are discharged into the atmosphere. They are absorbed through the skin and when we breathe the air in which there are these molecules, they enter our lungs. They are then carried by the blood to all the parts of the body, and stimulate the immune system, thus bringing about healing.[88] Each essential oil has its own particular aroma, and its own properties as well as its own benefits. Aromatherapy diffusers are scented with pure essential oils, creating a very soothing aura. If they are scented, they create an atmosphere that is calm and tranquil.

Aromas that are generally used are–

- Lavender, Pine, Chamomile, Peppermint, Jasmine and Orange relax the body and mind.
- Lemon, Rosemary, Sandalwood, Cedar wood, and Myrrh stimulate the body and mind and are energizing.
- Aloe and Lavender relieve painful swellings.
- Ylang Ylang and Chamomile help counter depression and high blood pressure.
- Geranium, Marjoram and Clary Sage create a breath of fresh country air.

Take a nice deep breath the next time you go to your dentist's office. You will know if he uses any aromatherapy. If not, you are likely to smell drilled teeth… yuck.

CHAPTER 7

FACTORS 61–80
TOP 20 SERVICES YOUR DENTIST
SHOULD PROVIDE

Factor #61: Regular Dental Cleanings

The routine dental cleaning is the basic maintenance that just about every dentist offers. When teeth are not cleaned regularly, a sticky bacteria-infested film gets deposited over the teeth. This is called plaque. When plaque is not removed regularly, it hardens to form calculus or tartar.[89] Also, the saliva in our mouth contains Calcium, and over time because of irregular cleaning, the Calcium and other minerals deposit on the surface of the teeth. While this is usually of the same color as the teeth, it goes undetected, but sometimes it may be brown or black, in which case the teeth look discolored.

Reasons why teeth need to be cleaned regularly are for the prevention of:

- Dental caries/cavities
- Periodontitis
- Gingivitis
- Oral cancer

While it is necessary to get the teeth cleaned by a dentist or hygienist, it is both important and necessary to clean the teeth regularly at home as part of a total oral hygiene program.

At home, teeth should be cleaned at least twice a day, and preferably after every meal. Use a slightly angled toothbrush with nylon bristles and of medium texture. The toothbrush should be replaced every month. It is best to use fluoride toothpaste. Flossing is essential at least once a day. This is the only way that the area between the teeth can be kept cleaned.

Professional cleaning of the teeth or prophylaxis—*prophylaxis* is a Greek word which means *to prevent beforehand.* Very, significant, too, since keeping the teeth clean will prevent periodontal disease.[90] This is usually done by the dental hygienist. The method used is:

- *Ultrasonic instruments*—These instruments use the vibrations produced by sound waves to remove tartar. As the tartar is removed, a spray of water simultaneously washes the debris and irrigates the area.
- *Hand-held tools*—Scalers and curettes are used to remove small deposits of tarter and to smooth the surface of the teeth.
- *Polishing*—Once the surface of the teeth is smooth, a slow-speed hand piece with a soft rubber cup that spins is used to polish the teeth. Prophylaxis paste is used for this.
- *Fluoride*—This helps prevent decay on teeth and around fillings and crowns.

It is best not to eat or drink anything for about an hour after the cleaning is done. It is important to have the teeth cleaned professionally at least twice annually. Make sure your dentist offers regular dental cleanings when choosing the right dentist for you.

Factor #62: Laser Bacterial Reduction

Periodontal diseases are bacterial diseases of the gums and teeth. The infection can destroy the bone and gums around the teeth. The trouble is that diseases of the gums are painless, and so go undetected till it is quite advanced. Periodontal disease causes bad breath (halitosis), swollen and bleeding gums, and eventually loss of teeth.[91]

Laser beams are tiny beams of concentrated light. When laser beams are directed on the infected area, it completely removes the infection, which is the cause of gum disease.[92] Laser beams work only on the area they are focused on, consequently, they are precise and accurate, and do not affect the surrounding areas. Laser beams sterilize the area and seal the blood vessels, thus completely stopping the bleeding and eliminating any chance of the infection spreading.[93] It is painless and there is no danger of any swelling or discomfort. Laser therapy needs only minimal anesthetic, though the dose is increased for patients who are unduly stressed or anxious. Laser beams used along with scaling and root planing increases the effectiveness of periodontal therapy.[94]

The point to remember is that laser beams have different wave lengths, and power levels. It is important to know which kind of laser beams are effective for which procedure so that there is no danger of damage to the tissues that are being treated.

A healthy measurement between your teeth and gums should be 1–3mm (called a periodontal pocket). If this pocket is 4mm or deeper, the bacterial count is too virulent, or too high, for your body's natural defense mechanism.[95] Laser Bacterial Reduction (LBR) is the best solution to stop the bleeding and infection when your pockets are 1–3mm in addition to your oral hygiene care at home. LBR helps reduce bleeding and promotes healing of the gums when gum disease is in its early stages.[96] You want LBR if your gums bleed when you brush or floss.

Factor #63: Deep Laser Gum Therapy

If you have pockets that are 4mm or greater, you have periodontal disease, or gum disease. To treat this disease, you have two choices to address it. Choice #1: Traditional periodontal surgery, where a periodontist cuts your gums open, cleans your teeth, and stitches you back up. Choice #2: Deep laser gum therapy.

Deep laser gum therapy (DLGT) is done using a dental laser. Once the teeth and gums are made numb, a small laser fiber is put into the space between the teeth and the gums. The energy from the laser beam kills the bacteria and accurately removes only the dead tissue. An ultrasonic cleaning instrument then removes the tartar from the surface of the teeth and roots, simultaneously washing away the debris. The laser beam is once again passed around the gum tissue sealing it so that there is no danger of any further infection. It has been found that 98% of patients who undergo laser periodontal therapy do not show any progression of the disease after 5 years.[97] Dentists or dental hygienists who perform DLGT need to be trained in the use of this laser.[98]

Thus, the benefits of DLGT are:

- Quick recovery
- Minimal risk of infection
- Minimal bleeding
- Greatly reduced discomfort post-treatment

In some areas where pocketing is extremely deep, your dentist may need to adjunct DLGT with antibiotic periodontal therapy. For antibiotic periodontal therapy to be really effective, it must be done along with DLGT, along with continued excellent oral hygiene. The number one technique is:

- *Arestin*—this is Minocycline but looks like powder. After the DLGT is done, Arestin is put inside the deep periodontal pockets. Arestin is a sustained-release drug and the medicine is slowly released by the process of polymer hydrolysis.[99] The antibiotic is then absorbed into the gum tissue where it kills the bacteria. Arestin remains inside the pocket for 21 days, during which time it effectively works on the diseased gum tissue. Arestin is the most popular antibiotic periodontal therapy used by dentists, and is especially effective in cases where the infection has created a deep pocket. The Arestin brings about a shrinking of the pocket, and heals the gums. Dental care givers can easily administer this antibiotic. Arestin has hardly any negative side-effects, and is considered to be a very safe drug. However, Arestin should not be given to children, pregnant or nursing women. It should also be remembered that after the dental care giver has administered the Arestin, the area should not be touched. Brushing of teeth should happen only after 12 hours have lapsed after the treatment. Flossing of teeth or using toothpicks should be done only after 10 days. The patient should avoid eating hard, sticky, or crunchy foods.

Follow-up dental visits should happen regularly so that the disease does not recur. If you have periodontal disease, ask your dentist about deep laser gum therapy and Arestin.

Factor #64: Oral Cancer Screening

Oral cancer mainly affects those who use tobacco, either alone or with alcohol, or have long periods of exposure to the sun.[100] A diet high in fruits and vegetables helps prevent oral cancer.

Oral cancer can be detected at an early stage, and treated successfully. Oral cancer often starts innocuously as a tiny red or white spot in the inner lining of the cheeks, lips, palate, gum tissue or the tongue. Other signs of oral cancer are: mouth sores that bleed easily or maybe do not heal, a change in the normal color of the oral tissues, a lump or thickening, a rough spot or crust, pain or numbness, difficulty in chewing, swallowing or when moving the jaw or tongue, or a change in the alignment of the teeth. A screening enables the dentist to discover these signs.

The best oral screening device is the Velscope system, which uses the technology of fluorescence visualization.[101] This is done as part of the usual oral examination. It is easy to administer, and takes only 2 to 3 minutes. There is no pain involved, no discomfort, and it is not messy.

How the procedure works:

- First, the dentist uses the *white light* to see if there are any lesions or abnormal tissues in the mouth. He also palpates the neck and face to check for any lumps.
- Using the Velscope handpiece, your dentist examines the oral cavity. The handpiece gives out a blue fluorescent light which is absolutely safe. This light is similar to that which a pulmonary physician uses to evaluate the lungs for cancer. The light penetrates and illuminates all the tissues from the top of the gum to its base. The special optical filtering in the hand piece helps the dentist distinguish changes in the light. If there is no problem, the tissues look candy-apple green, but if there is any abnormality, the fluorescence shows out the abnormal tissues as very dark irregular areas. This examination takes only a few minutes.
- Should your dentist see something abnormal, or if he has a suspicion as regards some wound in the mouth, he can ask for a surgical biopsy to be done. In such an event, his diagnosis will be based on the report of the oral pathologist.

Ask your dentist if he does oral cancer screenings. Then ask if he uses any additional devices to aid him do it right.

Factor #65: Comprehensive Dental Physicals

Comprehensive dental physicals are imperative for good health. It should be done if a comprehensive treatment plan is needed. Comprehensive dental physicals help in the detection of dental problems that with timely treatment can be cured.

A comprehensive dental physical includes:

- An interview: Establishing rapport between you and the dentist to see if you are right for each other. When you know that the dentist has your best interests at heart, the whole experience is worthwhile for both you and the dentist.
- Reviewing of the medical and oral history of the patient. This includes any previous surgeries, health concerns, or medication that you might currently be on.
- Getting a CT scan and any additional x-rays of the mouth done in order to get a proper and full view of the teeth as well as the condition of the bone.
- Taking impressions of the teeth to check the alignment, the bite, the condition of the teeth, and to look at the jaws in three dimensions.
- Checking the teeth individually for signs of decay, erosion, or build-up of plaque. The dentist reviews any treatment that the patient has undergone, including any restoration that might have been done. He meticulously checks the current condition of the teeth to see if they are chipped, cracked, or if there is some problem that needs attention.
- Oral cancer screening. This is done both visually and digitally to check for cancer. A thorough examination with a Velscope examination of the mouth, insides of the lips and cheeks, the hard and soft palate, the floor of the mouth and the tongue are done. The neck and face are examined by palpating and probing the areas concerned. The dentist also checks the lymph nodes. During the

examination of the face, the dentist checks for symmetry, swelling, jaw movement, and twitching.

- The extra-oral examination also includes checking the extra-oral muscles of mastication.
- Intra-oral examination includes examination of the intra-oral muscles of mastication. It also involves the examination of the oropharynx and tonsils.
- A periodontal examination. The dentist checks for signs of any gum disease.
- An occlusal examination. During this examination, the dentist inspects the teeth for alignment problems, worn teeth, and problems of the jaw such as TMJ or Temporomandibular joint disorder.

Regular comprehensive dental physicals help the dentist to review the regimen of oral health care. Patients are also educated on what symptoms to look for which may be suspicious, and which need immediate dental care.

Factor #66: Crowns and Bridges

Dental crowns and bridges are a part of restorative dentistry. Crowns and bridges can last for a very long time, with good oral hygiene. If a person has a crown or a bridge, care should be taken not to eat or drink anything very hot or cold, and certainly not to chew on hard foods.

Crowns—these are used to either cover an entire tooth, or to cap it, in case it is damaged. Crowns are used for endodontically treated teeth, or to improve a smile.[102] While it primarily strengthens the tooth, it also makes the tooth regain its shape, and consequently improves alignment. Most important is that it restores the confidence of the patient concerned. Crowns can also be fitted on dental implants. Crowns can be made of porcelain, acrylic, gold or alloys of certain metals. Sometimes, a combination of porcelain and a metal is used. Ceramic crowns are of different kinds: all-ceramic crowns, leucite-reinforced ceramic (Empress), ceramic with glass (In-Ceram), and alumina and zirconia ceramic (Procera).

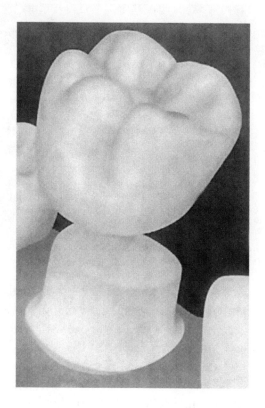

Bridges—these are one option to be used when there are teeth missing. If there are gaps between the teeth, the remaining teeth tend to shift or move into the empty spaces. Because of the imbalance, gum disease can develop, or there can be temporomandibular joint (TMJ) disorders. Bridges are cemented to the teeth or implants that are on either side of the gap, known as abutments. The replacement tooth or the pontic, a component of the dental bridge, is fixed to the crowns that cover the abutments. Bridges are made of porcelain, or ceramic. The dentist decides which kind to use depending on where the bridge is to be fixed, and the function it will have to perform.

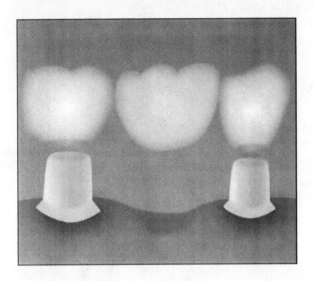

Before the crown or bridge is fixed, the teeth have to be prepared. This involves removing the enamel or a part of the teeth. The taper, margin and ferrule effect will have to be carefully worked out. The color of the crown or bridge is matched with the other teeth. CAD/CAM technology helps the dentist prepare the crown or bridge with the greatest possible accuracy and precision. Make sure your dentist can do crowns and bridges.

Factor #67: Root Canal Therapy

The tooth has two basic parts: the crown and the root. Inside the whole tooth is dental pulp which is a soft tissue containing nerves and blood vessels. The blood vessels carry the nutrients, and the nerves provide the sense of feeling. In case there is a cavity, a loose filling, or a crack, bacteria enter the pulp, and destroy it. This infection goes deep down into the root canal and the tooth becomes extremely sensitive to hot and cold foods, causing great pain. If not treated, the bacteria go deeper through the root openings and spread the infection to the bone below. Once the bone is infected, there is every possibility that the ligaments which attach the roots to the bone swell and loosen the grip on the tooth.[103]

Root canal therapy is a modern endodontic procedure which aims at fixing the damaged tooth. During this therapy, first, the tooth is made numb. In case there is any existing filling, it is cleaned out. Any infected pulp is removed. The root canal system is then cleaned, enlarged, and shaped to make it ready to be filled with a suitable permanent filling. Medicines are put into the pulp chamber and canal to ensure that all the germs are killed and there is no remnant infection.[104] In order to ensure that the tooth is completely cleaned and there is absolutely no bacteria present, the patient will have to go for several sittings. Sometimes, so that no infection starts between visits, the dentist might put a temporary filling in the crown. The dentist may prescribe medicines to prevent the infection from spreading. Once the dentist is sure that there is no infection, and the canal and pulp chamber are absolutely clean, he will put in the permanent filling. Then, the crown is constructed and the tooth restored.

A tooth which has undergone root canal therapy tends to be brittle. Therefore, a tooth that has been root canal treated requires a crown to give it strength and support.

Most dentists will refer molars (back teeth) to dentists who specialize in root canal therapy. These specialists are called endodontists. They use microscopes to visualize the canals. Often times upper molars can have 4–6 canals and lower molars have 3–5 canals. Without a microscope, it is impossible to properly clean these canals. Make sure your dentist does root canal therapy and refers the tough cases to a specialist he trusts.

Factor #68: Extractions

A dental extraction is the removal of a tooth. There are many reasons why a tooth needs to be extracted.[105] These include:

- Severe infection or decay
- Severe gum disease
- Fractured teeth
- Disease in the impacted 3rd molars or wisdom teeth, or in the area near the 3rd molars
- Severe crowding
- When a complete denture is required
- In case of radiation therapy for the head and neck

A dental examination and x-ray give the dentist a clear picture of the tooth, and help him decide what kind of extraction will be needed. Preventive antibiotics may be given to patients who have a high risk of infection.

Post-extraction, the patient has to bite on a piece of sterile gauze for 45 minutes.[106] However, bleeding is normal for about a day or so. There may be pain, swelling or stiffness in the jaw, which eases after a couple of days. Cold compress, pain killers, and rest help the patient cope. The mouth needs to be rinsed gently with warm salt water. To prevent dry socket, patients are advised not to spit out the saliva.

Factor #69: Dental Implants

Swedish professor Per-Ingvar Brånemark discovered, in 1952, that when titanium is placed into bone, osteoblasts (bone cells) grow on as well as into the implanted titanium, forming a structural and functional bond between the two.[107] A dental implant is an artificial tooth replacement made of titanium and is threaded like a screw.

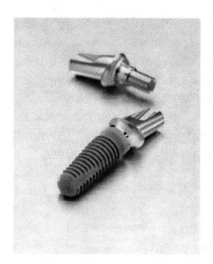

Dental implants can be placed by a dentist, implant specialist, oral surgeon, periodontist, prosthodontist, or endodontist.

There are multiple reasons to place dental implants.[108] These include:

- *Aesthetics*—they look like natural teeth
- *Comfort*—since there are no more gaps
- *Durability*—they bond to the jaw bone, becoming a part of it
- *Improved oral health*—other teeth remain intact, and the bone area is stimulated
- *Improved hygiene*—allows for normal dental care
- *Replacement of a single tooth*—is possible

The procedure involves the following phases:

- *Planning*—A CT scan will need to be taken. In some cases, a 3-D CAD/CAM may be used to plan the procedure
- *Surgery*—the implants are fixed into the jaw bone
- *Restoration*—the crown is built and attached

The duration of the whole procedure depends on the patient's physical condition, medical history, and anatomy. After the surgery, it takes 2 to 6 months for the implant to fuse with the jawbone. The next step involves uncovering the implant to fix the extension. For the restoration, the crown is built and attached to the abutment of the implant. In some cases, the dentist might do the restoration work immediately after the implant surgery. This procedure is known as *Immediate Loading*.[109]

In case several teeth have to be replaced, implant-supported bridges are used. This is multiple implants with multiple teeth attached. Sinus augmentation and ridge modification, if required, will increase the chance for successful dental implant integration. The All-on-Four technique developed by Dr. Malo allows someone who has no teeth in a jaw to get a full set of fixed teeth with as little as four implants.[110]

In order to get the best level of care, make sure your dentist is affiliated with the American Dental Implant Association and the International Congress of Oral Implantology. These dentists have had extensive post-graduate education on implant dentistry. If they have fellowships, masters, or a diplomat from these institutions, you can trust that they have been comprehensively trained and tested in the field of implant dentistry. Ask your dentist if he places and restores dental implants. If so, make sure he uses a reputable implant such as CAMLOG, Nobel, or Zimmer.

Factor #70: Bone Grafting

Bone grafting is needed if you do not have enough bone mass for a dental implant.[111] Bone loss may be due to a previous dental extraction, periodontal disease, trauma, or worn dentures.[112]

There are four types of bone grafting materials available.[113] These types of bone graft are:

- *Autograft*—Bone from you! The bone is usually harvested from your mouth, hip, or tibia. Since the bone is from your body, there is no risk of disease. The graft will not be rejected, and the new bone grows easily.
- *Allograft*—Bone from someone else. Bone from a donor is used. Bones taken from a cadaver are put through many tests and processes to become sterile. It is 100% safe and less painful than an autograft.
- *Alloplastic graft*—Synthetic material is used. This is made in a lab. It is a kind of calcium phosphate and encourages bone growth. This graft gets gradually replaced by your natural bone. Some materials, though, form a kind of scaffold on top of which the bone can be built. All materials used in this kind of graft are bio-degradable, and completely safe.
- *Xenograft*—Bovine (cow) bone is used. The bone is processed to sterilize it and make it bio-compatible. Eventually this gets replaced by natural bone from the patient's body.

Along with the bone graft, protective membranes, called barrier membranes are grafted on, in order to stabilize the bone. He also prescribes an antibiotic mouthwash. The dentist x-rays the area frequently to check on the progress of the bone graft. When the height and width have reached the required proportion, and the graft is completely healed, the area is ready for the dental implant.

Make sure that your dentist can do bone grafting if he places and restores dental implants.

Factor #71: CT Scans & Digital X-Rays

Cone beam computed tomography (CBCT) has enhanced the ability to identify findings and pathology within the oral and maxillofacial complex. In a short period of time and with far less radiation than medical CT, CBCT produces high-resolution three-dimensional data. A computer compiles all these images into complete cross-sectional pictures, so that the bone, teeth, and tissues are all clearly visible.[114] After reviewing the multiplanar reconstructed images, the dentist can communicate findings with you, the patient. It is important for the dentist to spend the necessary time to record and document findings as well as manipulate the data for proper treatment planning. If done correctly, treatment planning and executing surgical and restorative needs for the patient will result in expectations met by the practitioner and patient.

The CBCT machine has a small footprint therefore requiring space not much greater than that occupied by a panoramic imaging machine. In one rotation of the source and receptor, the receptor captures the entire anatomical area, called the field of view. With no magnification, distortion, or overlap of anatomy, measuring is precise as long as the anatomical area is oriented correctly.

The information obtained from the data from a 3-dimensional CT scan offers many advantages.[115] These include:

- The images are 3-dimensional and highly detailed
- An exposure of 9–24 seconds is needed
- More accurate than a medical CT scanner
- Help in implant and bone grafting treatment planning and follow-up
- Allow the dentist to view and analyze the temporomandibular joints
- Gives a great depiction of impactions and lesions which are not visible during a traditional set of dental x-rays
- Low radiation dosage: 50 times less radiation when compared to a medical CT scanner
- The dentist can do a virtual dental implant surgery before the actual surgery
- Make airway studies
- Evaluate any anatomic variations and pathology

- Evaluate injuries which have resulted from trauma

The data obtained helps in planning surgery for implants, orthodontic assessments for impacted teeth and delayed tooth eruptions in relation to the adjacent teeth, orthognathic oral surgery, 3-D cephalometric analysis, sleep apnea dentistry, TMJ evaluation, and in endodontics.

The CT scan is particularly useful for dental implants. The picture is so accurate, that the dentist can make an exact 3D copy of the jawbone from the data that he gets.[116] Third-party software incorporates CAD/CAM technology to make replicas of dental implants that can be placed within the upper jaw (maxilla) and/or lower jaw (mandible) after which a surgical guide can be fabricated to deliver dental implants with precision. The future of dentistry lies in digital imaging using DICOM data. In addition, using peer-to-peer software (i.e. GoToMeeting by Citrix) allows dentists to communicate without leaving their offices.

CT Scans show dentists the big picture. They are costly, so often times, digital dental x-rays are indicated. A digital dental x-ray helps the dentist diagnose individual problems in the teeth and jaws. When using digital dental x-rays, a flat electronic pad or sensor collects the data. The x-rays hit this sensor and the image is directly sent to the computer, where it shows up on the screen.[117] The image is stored for further use, or it can be printed. This is of advantage when either the dentist or the patient wants a second opinion. Another great advantage is that since the image is stored, it can be compared with previous and subsequent x-rays, and the progress of the patient can be clearly mapped.

These x-rays show details that are not visible to the eye, such as early signs of tooth decay that might be between the teeth, or under a filling, cracks in an existing filling, infection in the gums, problems in the root canal, or any other abnormalities in the mouth.[118] They also help the dentist initially plan implants or any other dental work. In fact, when discussing the problem with the patient it is possible to zoom into the exact area so that the patient can see clearly what the problem is.

Ask your dentist if uses digital x-rays and CT scanning.

Factor #72: Composites

Do you want mercury in your mouth? No! Composites are tooth-colored filling materials that can be used when you have small cavities or small chips in teeth. A composite is made up of filler particles surrounded by a matrix. The matrix binds the filler particles together. Filler particles are of varying coarseness, while the matrix is either a paste or a liquid, which hardens when activated. Activation is done by a catalyst (which is added with the filler particles), or water, or some other solvent. Before hardening, it is put into a mold.

Dental composites are:

- The most commonly-used composite resin is made up of a filler of finely-ground glass particles and an acrylic matrix also called BIS-GMA (Bisphenol A-Glycidyl methacrylate) or UDMA (Urethane dimethacrylate).[119] To ensure bonding of the filler and the matrix, the filler particles are coated with a silane-coupling agent. In case a photo initiator is used, the catalyst is mixed with the paste which becomes active when a strong light beams on it. The acrylic matrix hardens when the catalyst starts functioning.

- Dental amalgams (silver fillings) are made of finely ground silver or tin, mixed with mercury: a known neurotoxin.[120]

Direct dental composites—this is done by the dentist. A hand piece with a curing light is held as close to the resin surface. This light emits wavelengths that are specific to the initiator and catalyst involved. Ensure that the eyes are protected from the curing light. Direct dental composites are used for filling gaps between teeth, making partial crowns on single teeth, or for the minor reshaping of a tooth.[121]

The main advantage of using composites is aesthetics. Composites can be made such that the color is as close to the natural color of the teeth, making for almost invisible restoration. Composites strengthen the tooth as well. For a dentist, using composites is very convenient. He can take his time to shape the composite till he is satisfied with it, and then expose it to the curing light. Dental composites harden within 7 to 9 seconds of being exposed to the curing light. Make sure your dentist does not do amalgams and does do composites.

Factor #73: Whitening

Who wants whiter teeth? Ever since *Extreme Makeover*, whitening has become a popular treatment. Dental whitening or bleaching is a component of cosmetic dentistry. As people age, their teeth slowly lose their whiteness. The enamel becomes stained due to various reasons that include having too much tea, coffee and cola, smoking, and certain kinds of medication.[122]

Though a dilute solution of hydrogen peroxide is used to bleach teeth, it has been found that using 10% or 16% carbamide peroxide solution is the safest form of bleaching.[123] It does not burn the mouth, or damage the teeth. However, bleaching solutions temporarily may cause the teeth to become sensitive. Fluoride dramatically reduces this sensitivity.

Whitening toothpastes have a low percentage of carbamide peroxide, and does not produce dramatic results.[124] Safe over-the-counter bleaching agents should have the ADA seal.

Bleaching strips are thin plastic strips that stick to the teeth. The concentration of carbamide peroxide in these strips is between 7% to 14%. They are worn for 30 minutes twice per day.

Bleaching trays on the other hand are quite effective. This process is done by the dentist. The bleaching agent is carbamide peroxide and is used in concentrations of 10% to 35%. Impressions of the upper and lower teeth are taken, so that the bleaching trays are custom-made to fit. A drop of the bleaching agent is put in each tooth indent, and the tray is put in the mouth with the teeth fitting snugly into the indents. The trays are worn for 30 minutes to an hour. The carbamide peroxide penetrates through the enamel to the dentin below, bleaching it.

Bleaching gels are a quick way of bleaching in the dental office. The carbamide peroxide gel is placed on the teeth, and is activated with a light source. The strength of the gel is determined by the dentist according to the amount of discoloration there is. The light activates the crystals in the gel, which absorb the light energy, and penetrate the enamel of the teeth.[125]

Ask your dentist if he whitens teeth, and then ask him which techniques he uses.

Factor #74: Veneers

Do you want that Hollywood smile? Veneers can make that dream become a reality. Veneers are restorative materials. They are very thin (0.3mm to 0.5mm) and are custom-made to fit on the front surface of the teeth.[126] They change the shape, color, symmetry and size of the teeth.

People go in for veneers for various reasons.[127] These include:

- To fix teeth that are worn, chipped or broken
- To permanently change discolored teeth
- To fix teeth that are uneven, irregularly shaped, or not aligned properly
- To close gaps between teeth
- To improve a smile

Veneers can sometimes be prepared without anesthetic. For those who have any anxiety, a small dose of anesthetic is given. A 0.5mm reduction of the surface of the tooth may be necessary, so that the veneer fits well and does not look bulky.

A point to be remembered is that if there is serious discoloration, it might not be possible to match the exact color. In this case, multiple veneers must be used.

A dental examination is necessary to make sure there is no underlying gum disease (periodontitis). The dentist will need x-rays of the tooth or teeth involves. After drilling, an impression is taken and sent to the laboratory for the construction of the veneer. The veneer is then tried on the tooth to see if it fits. Adjustments, if any, are made to the tooth, as well as to the veneer. After the veneer is ready, the tooth is cleaned, polished and prepared for cementation. Special cement is spread on the veneer, and then it is placed on the tooth. Next, a special light is beamed on the veneer, which activates the molecules of the chemicals in the cement causing the cement to set or cure.[128] Excess cement is removed and final adjustments made. Voila! You have a new smile! Ask your dentist about veneers.

Factor #75: Dentures

Natural teeth may be lost due to periodontal disease, tooth decay, or injury.[129] Dentures are prosthetic restorations which replace a full set of upper and/or lower teeth when they are lost.

Dentures have many benefits.[130] They can help in the following ways:

- *Aesthetics*—they restore the appearance
- *Mastication*—they help in chewing the food properly, consequently aid digestion
- *Phonetics*—they help in enunciating words clearly
- *Self-esteem*—they help patients feel good about themselves

In order to get dentures, a few steps need to occur.[131] The first step is a comprehensive examination including a CT scan. Impressions of the upper and lower ridges are taken. If the patient has an existing set of dentures, those are replicated and relined.

Plaster is poured into these impressions to make models. The shade, size and shape of the teeth to be placed are determined. The first set of impressions is used to make a second set of impressions which have a perfect fit. Factors like the length of the teeth, plane of the teeth (i.e., the teeth should be parallel to a line between the pupils of the eyes), and the alignment (so that when the patient bites, the upper and lower teeth are perfectly aligned) are worked out.[132] These teeth should be visible just below the lips. Speech also needs to be tested. Then the laboratory places teeth into a fixed position. Any adjustments that have to be made are done at this point. When the patient and the dentist are fully satisfied, the denture is sent to the lab for processing and finishing.

Types of dentures:

- *Standard dentures*—the back of the denture ends behind the hard bone which is in the roof of the mouth. If this is uncomfortable, then the back of the denture is removed. This does make the denture unstable.
- *Immediate dentures*—these are also called temporary dentures and are made before any natural teeth are removed. Once the teeth are extracted, these dentures are fitted immediately. The lab estimates where the gums will heal, so this is not a long term prosthesis.
- *Implant retained dentures*—these are also used quite effectively, especially for those who cannot wear lower dentures.

Ask your dentist about dentures.

Factor #76: Partial Dentures

Partial dentures are used when a patient has lost only a few teeth. In fact, partial dentures are important in keeping the remaining teeth in place, so that they do not move into the gaps, and damage the oral tissue. Further, it is difficult to clean the teeth properly when there are spaces in between. This might lead to periodontal disease, and tooth decay.[133] Fitting partial dentures have other advantages: it will enable the patient to chew his food properly, speak clearly, and restore his smile.

A partial denture is removable. It is made up of replacement teeth that are fixed to a pink plastic base. This is attached to the remaining natural teeth in the mouth either with metal clasps or precision attachments. Crowns on the natural teeth help the partial denture to fit better.[134] The patient should be able to remove the denture or put it in, with ease, after a little practice. Partial dentures should never be forced into place. Besides breaking the clasps, there is every possibility of damaging the mouth and the anchor teeth.[135]

Some precautions that need to be observed are:

- Patients with partial dentures need to maintain their oral hygiene extremely well, or future dental issues will develop
- In case of pain or discomfort, the patient should immediately contact the dentist.
- Only soft food and small bites should be had for a while after the partial dentures have been fitted. Completely avoid sticky and hard foods, including chewing gum.
- Food should be chewed by both sides, so that there is even pressure on the denture.
- Reading aloud to practice speaking clearly.

Other forms of partial dentures are:

- *Cast metal partial dentures*, which are especially good for patients with TMJ disorders. One drawback is the esthetics: metal clasps don't look great if you have a high smile line (in terms of where your lips go when you smile).
- *Valplast* and *Flexite* (made of nylon). The clasps are the color of gums making them almost camouflage with the rest of the mouth.

- *Vitallium-Valplast*, which is a combination of cast metal with Valplast clasps.

Ask your dentist if partial dentures are right for you.

Factor #77: Sedation

Oral sedation is given to nervous and frightened patients, preferably the night before the dental procedure is scheduled, so that they are calm when they come to the dental surgery, and have a pain-free experience. Common forms of sedation are: inhalation of Nitrous oxide; oral sedation with Benzodiazepines such as Halcion (triazolam) or Ativan (lorazepam); and Intravenous conscious sedation with Midazolam and Diazepam (though barbiturates, opioids, and Propofol can be used).[136]

Benzodiazepines are known as Central Nervous System depressants, because they bind with the receptors in the brain which in turn act on the parts of the brain that produce the emotion of fear.[137] Benzodiazepines have a sedative-hypnotic effect, which make the patient drowsy, and therefore calm. [138] When given in higher doses, they induce a kind of sleep or hypnotic condition.

Naturally, the dose is prescribed by the dentist, since he is fully aware of the dental work that needs to be done as well as the psychological condition of the patient.

For dental procedures, the safest forms of sedation for adults are Nitrous Oxide and oral Benzodiazepines.[139] Benzodiazepines used as sedative-hypnotics are:

- Halcion (triazolam)—0.25mg to 0.5mg given 1 hour before bedtime, and 1 hour before the treatment.
- Ativan (lorazepam)—2mg to 4mg given 1 hour before bedtime, and 1 hour before treatment.

Oral conscious sedation using Triazolam with local anesthesia or a combination of Triazolam with both local anesthesia and Nitrous oxide has been found to be very effective.[140]

Points that the patient should be reminded about for the rest of the day are:

- To have an escort drive you to and from the dental office
- To avoid strenuous activity
- To avoid eating after midnight the day before treatment
- To wear loose clothing

If you want to be put to sleep during your dental visit, ask your dentist if he offers sedation.

Factor #78: Nightguards/TMJ Therapy

Do you grind or clench your teeth? Does your jaw joint hurt? Do you ever wake up with headaches? If you answered yes to any of these, you may need a night guard. A night guard is made of moldable plastic and comfortably fits on top of the upper or lower teeth. It acts like a cushion between the upper and lower teeth. While some night guards can be bought over-the-counter, some are fabricated to get a good fit. The material used to make night guards is polyester thermoplastic nylon hard acrylic and EVA or ethylene vinyl acetate. Night guards are used for bruxism (teeth grinding and teeth clenching) and TMJ disorders.[141]

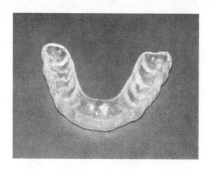

The TMJ or temporomandibular joint is found on either side of the head, just in front of the ear. It is the point where the maxilla (upper jaw) and mandible (lower jaw) meet. This joint is a ball and socket joint with a disc in between. The TMJ is made up of muscles, tendons and bones, and is used in biting, chewing, yawning, and talking.

TMJ disorders are brought on by bruxism, nail biting, chewing of gum, misalignment of the teeth, trauma to the jaws and stress.[142] Problems in the TMJ lead to stiffness, headache, earache, tinnitus (ringing in the ears), dizziness, and problems with biting. Proper diagnosis of TMJ disorders requires an excellent clinical examination along with a cone beam CT scan.[143]

TMJ therapy has various treatment modalities.[144] These include:

- *Rest for the jaws*—avoiding hard and chewy foods, and not opening the mouth very wide.
- *Heat and ice therapy*—immediately after an injury to the TMJ, cold packs are best. However, if the pain continues, alternating hot and cold packs relieves muscle spasms.
- *Medication*—anti-inflammatory drugs, for instance, aspirin, naproxen, steroids or ibuprofen, control inflammation. Diazepam relaxes the muscles. If the situation is serious enough to merit it, then methylprednisolone and triamcinolone, which are cortisone preparations, can be injected into the TMJ.
- *Physical therapy*—massaging or giving an electrical stimulation help decrease the pain. Slowly opening and closing the jaw will strengthen the joint and ease movement.
- *Stress management*—support groups, counseling, and medication help the patient realize that he is not struggling alone.
- *Corrective dental procedures*—Dental restoration or orthodontics may be needed to correct or improve the

bite. If the patient has crowns or bridges, adjustments will need to be made so that the teeth are properly aligned.

- *Occlusal therapy*—this is the use of night guards. They may need to be worn during the day as well. One exceptional type of night guard for people who clench is the NTI-TSS, which goes over your front 2—4 teeth.
- *Surgery*—when all else has failed, surgical procedures such as joint-restructuring, joint replacement, ligament tightening, or TMJ arthroscopy will need to be done.

Talk to your dentist if you are clenching your teeth, grinding them, or experiencing TMJ problems.

Factor #79: Sleep Apnea Appliances

Do you snore? Does your spouse or loved-on snore? If so, there may be sleep apnea present. Sleep apnea is a very serious issue. It actually causes people to stop breathing during sleep. These pauses in breathing can occur along with shallow breaths while sleeping from 20 to 30 times or more per hour. Normal breathing resumes with a choking sound or a loud snort.[145]

Symptoms of sleep apnea include morning headaches or migraines, excessive sleepiness in the daytime, short-term memory loss, slow metabolism, difficulty in concentration, and insomnia.[146]

Problems that arise directly because of sleep apnea are hypertension, heart attack, stroke, obesity, and increased chances of car accidents.[147]

There are 2 kinds of sleep apnea—Obstructive sleep apnea, which is the most common, and Central sleep apnea when the brain does not send signals to the breathing muscles.[148]

Diagnosis of sleep apnea is done using a polysomnogram or PSG.

Treatment of sleep apnea requires the patient to bring about lifestyle changes, particularly in the diet. For serious cases, there are special appliances which can be used.

- *CPAP*—the Continuous Positive Airway Pressure or CPAP is an appliance to treat sleep apnea. The patient has to wear a special mask and tube which uses pressure to send air through the nasal passages.[149] This steady stream of air allows the patient to breath freely. The patient must wear this mask every night, and all night. CPAP cannot be used by those who suffer from allergies, or those who sleep on their stomach.
- *Oral appliances*—these are used to treat Obstructive Sleep Apnea. They prevent the obstruction of the airways. They are worn in the mouth while sleeping, and prevent the soft tissues from collapsing. These appliances reposition the lower jaw, tongue, soft palate and the hyoid bone such that the airways remain open.[150] Alternately, they offer muscle toning so that these muscles do not collapse. The FDA has approved dental sleep apnea appliances as long as these appliances are custom-fit and titrated for patients that have mild to moderate sleep apnea. The dentist will recommend the appliance which he feels would suit the patient best.

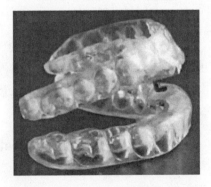

If all else fails, then the patient might need surgery to cure sleep apnea. Ask your dentist if he treats sleep apnea with oral appliances.

Factor #80: Free Initial Consultations

Going to a dentist is a daunting experience. It is very important for you to know and trust the dentist that you go to. Most dental care givers have their own websites which give details of their dental practice. This would help narrow down the choice to a certain extent. However, it is necessary to visit the dentist before you decide on whether he is the right choice. Most dentists, today, being sensitive to the feelings of patients, offer free initial consultations which are quite comprehensive.

At this time, you meet the team, meet the dentist, and are taken around the entire office. This gives you an idea of the atmosphere, and whether the latest in technological advances are balanced by humanness and ethics. You can talk to each member of the team, so that you know what each one does, and what type of treatments are offered. In a one-on-one informal conversation with the dentist, you would be able to explain your dental problem.

The dentist would talk about and explain the treatment that he thinks would work for you. Often, the dentist would offer to schedule a thorough dental physical: clinical examination of the teeth and gums, which would include the necessary CT scan and x-rays, digital photographs, inspection of the gums, muscle and jaw joint evaluation, and oral cancer screening. Since most of these can be shown on the computer, the dentist is able to clearly explain the problem and the diagnosis, and also outline the treatment that he thinks is best.

The dentist would offer all the available options, with their advantages and risks, as well as the costs involved, so that you can make an informed decision. The team might show you photographs of patients with similar problems who were treated successfully by them. The dentist would want to see previous records, both dental and medical, so that he could evaluate the current condition compared to earlier, and then outline a course of action.

In case the costs are high, the dentist would be able to advise you on financial resources that could be looked into and utilized. Ask your dentist if he provides free initial consultations.

Chapter 8

Factors 81–90
Top 10 Things a Doctor
Should Do Before Starting
Comprehensive Dentistry

Factor #81: A Thorough Exam

A dentist is not merely a tooth mechanic. It is important to have a thorough knowledge of dentistry, medicine, function, and all the latest technological devices that can be used to facilitate dental procedures. A comprehensive interview gives the dentist a lot of information regarding his patient's medical and dental history. The dentist needs to conduct an occlusal examination (an exam of the bite and function) along with a TMJ exam. [151]

You want a doctor who is motivated about his work, so that you know he will spend the right amount of time to diagnose what is present and to listen to you and find out what you really want. Then, he will explain and educate you about the issues at hand and talk about the various treatment options available to you. He will instruct you on the importance of and need for good oral hygiene. All this is done within the framework of the dentist's mission and philosophy. All this also takes a lot of time. As the scope of dentistry has increased, expectations of the patients has also increased. Communication, therefore, is of the essence. Ideal conditions are hard to find, and the only way to be happy in his work is for the dentist to work towards becoming an expert in technical skills as well as people skills.

You want a dentist who performs a thorough initial examination.

Factor #82: Periodontal Charting

This is a clinical examination done to evaluate the gums. The word periodontal means around the tooth. The gum tissue and bone should fit snugly around the tooth. When a patient complains of bleeding gums or bad breath, the dentist knows that the tissue and bone that supports the tooth is getting destroyed, creating a pocket around the tooth.[152] Left untreated, the pocket becomes deeper, providing copious space for bacteria to live in. These bacteria move on under the gum tissue, making the pockets even deeper. More bacteria come in and there is more loss of tissue and bone. Finally, the tooth becomes loose and either falls off, or has to be extracted.

A periodontal probe is used to find out the depth of the pocket, and consequently the seriousness and severity of the disease.[153]

New research has proved that there is a clear link between the inflammation caused by periodontal disease to diseases which are inflammatory conditions themselves. As the periodontal bacteria travel in the body, the immune system fights back with white blood cells. These WBCs release chemicals that create an inflammatory response as is found in diabetes, respiratory diseases, cardiovascular disease, rheumatoid arthritis, and Alzheimer's disease.[154] Therefore, when a patient is suffering from periodontal disease, it is best to consult a general physician along with the dentist.

You want to have periodontal charting during your dental examination to ensure that you are getting the best possible evaluation.

Factor #83: Oral Cancer Screening

The statistics released by the National Cancer Institute are alarming—every year about 29,000 people are detected with oral cancer, i.e., cancer of the mouth and lips, and oropharynx (part of the throat which is at the back of the mouth). The NCI provides the latest information on oral cancer, both on the internet (http://www.cancer.gov) and on the telephone (1–800–4-CANCER).

The most important thing is to go for regular check-ups. Early detection means a better survival rate.[155]

You are in the high-risk category if you:

- Are male
- Over 40 years of age
- Use tobacco in any form
- Drink alcoholic beverages once per day
- Have a history of head or neck cancer
- Have too much exposure to the sun (this causes lip cancer)

The symptoms include:

- A small white or red patch in the mouth
- A sore/lesion that bleeds easily and does not heal quickly especially on the floor of the mouth or on the base of the tongue
- Difficulty in chewing and swallowing food
- Loose teeth
- Lump in the neck

To detect the disease:

- The dentist palpates the side of the neck and under the jaw of the patient to check for lumps.
- Face, neck, lips and mouth are carefully inspected
- The tongue is checked for swelling, abnormal color or texture. The base and underside of the tongue are inspected.
- The roof and floor of the mouth and back of the throat are looked at.
- The Velscope examination

Make sure your dentist checks for oral cancer during your dental visit.

Factor #84: Cone Beam CT Scan

Cone beam computed tomography (CBCT) has dramatically improved the way your dentist evaluates your mouth. In a short period of time and with far less radiation than medical CT, CBCT produces high-resolution three-dimensional data.

In order for your dentist to properly evaluate your teeth, your jaw bones for dental implant placement and bone grafting, your TMJ, your airway, any anatomic variations, or previous injuries, a CT scan is needed.[156]

Make sure your dentist takes a cone beam CT scan prior to any major dental work.

Factor #85: Dental X-Rays

Dental x-rays are pictures of the mouth and teeth produced using a sensor or film. If there are silver fillings or metal restorations they block the photons and show up as dense white patches on the x-ray. Structures which contain air appear black. Teeth, tissues and fluid appear as shades of gray.

The types of x-rays are:

- Bitewing x-rays, where the patient bites down on a tab. This shows the crown portion and part of the root, of the top and bottom teeth, and the immediate surrounding bone level. Bitewing x-rays can detect dental cavities which may not be visible to the dentist, and so patients are recommended to take these x-rays every year.
- Periapical x-rays show one or two teeth right from the crown to the end of the root, as well as the bones that support the tooth. These show the presence of abscesses, and cysts and are also used to evaluate bone loss.

Ask your dentist to see these x-rays if you want to understand what is going on in your mouth.

Factor #86: Intraoral Photos

Most people need to see their dental condition to be able to understand and accept it. It is also very important for the dentist to be able to get a clear picture of the inside of the mouth of the patient. Using the intraoral camera, a technological marvel, the dentist can properly explain and educate you on good oral care.[157] This is a completely painless diagnostic process.

The intraoral camera is lightweight and portable, and the tiny sensor provides excellent detailed images of the mouth. It looks like a wand. There is a tiny video camera fitted to a handpiece. High quality LEDs help provide clear images.

The capture button located on the rear of the camera, makes it is easy to use. There are multiple mounting options which make for greater flexibility. Since it is connected directly to the computer through a high-speed USB, the images are instantly available. Together, the patient and dentist can discuss the dental problem and work out the treatment plan.

Dental problems that the intraoral camera makes visible include corroded or perculating fillings, bleeding gums, decay in the teeth, plaque, and fractured teeth. In case there is an incipient (or early) problem, this too is visible with the intraoral camera. These images can be stored, and they can be compared at every point of the treatment to check the progress of the treatment. These photographs are easily retrievable. Intraoral photographs make it possible to consult with other dental and medical personnel, if required.

Intraoral cameras come with disposable covers so that infections are not passed among the patients.

You are entitled to the best dental care possible, and using the intraoral camera to detect a dental problem is one way of ensuring that this happens.

Factor #87: Extraoral Photos

Digital cameras take excellent quality pictures, and the dentist can use these to build trust in his patients. Digital photographs are used to educate the patient. They can see their current profile, and by discussing this with the dentist and taking his advice, are willing to open their minds to the idea of having corrective procedures done.[158] At each examination, digital photographs taken build up a set of images so that the progress of restoration is clear. The images can be annotated, so discussing each stage becomes easier. Seeing digital photographs of patients who have had similar dental procedures allows patients to understand what they would need to undergo to achieve success.

Digital photographs are invaluable in case the patient has to be referred to other medical personnel for some treatment that may be required. A doctor receiving a file of digital photographs and x-rays, electronically, knows that the dentist is serious about his work and is up to date. The patient too feels confident about completing the treatment.

Affixing digital photographs of the patient, as well as digital photographs of x-rays taken, to insurance claims, make for a strong case. Since they are being sent electronically, they can be traced, and no claim is lost. Also, since these claims reach the insurance office in a short time, they are processed quickly. This, naturally, leaves more time for the doctor to give to his patients.

Good quality digital photographs of patients, who have been treated, when used on the website, make for authenticity. This becomes the dentist's own publicity.

If your dentist is taking photos, you can be assured that he cares about his work and he cares about you.

Factor #88: Impressions For Diagnostic Models

The dentist, as part of his diagnostic procedure, should create models of your teeth prior to any comprehensive care. Models are also known as diagnostic casts. This is a plaster mold of the upper and lower teeth. With diagnostic casts, a plan can be three-dimensionally constructed in wax. In fact, the wax re-creation of the teeth, also known as a wax-up, helps the dentist in outlining the specific goals you have to achieve.[159] The wax-up shows if there are any cosmetic problems that need attending to.

These will have to be corrected and treated before any implant work can be done. The dentist studies the photography, as well as the diagnostic cast to make decisions on what kind of dental implant would be best for you. He also decides on how to do the dental implant, working out, visually, the whole surgical procedure that he will have to do. In case an implant is required only for a few teeth that are missing, these diagnostic casts help to recreate the patient's missing teeth. Using this information, the dentist can plan out precisely where the dental implants should be placed.[160] Their positions and dimensions are worked out to the last detail.

Seeing the images, as well as the mold, makes the patient feel part of the whole procedure. By thus involving the patient, the dentist builds up the patient's trust.

The benefits of getting impressions are:

- Accuracy, since these images are magnified
- Reduction of time
- A complete picture
- Record keeping
- This is a less invasive method for getting data

You want diagnostic models created prior to starting any major dental care.

Factor #89: Bite Registration

Occlusion means the contact between the teeth and how a patient bites. Specifically, it is the relationship between the maxillary (upper) teeth and the mandibular (lower) teeth when at rest, and when chewing food. The bite registration tells the dentist exactly what the state of occlusion is. The common method used for bite registration is polyvinyl siloxane with the patient biting down.[161]

Most dental treatments involve the occlusal (biting) surfaces of the teeth, so it is important for dentists to know the movements of the mandible and the influences that control them. What affects bite registration are tooth wear, tooth movement, and fracture.[162]

A comprehensive examination of the masticatory system will need to be done. This includes the TMJ and the supporting muscles. In case there is any temporomandibular disorder, this will need to be attended to before any other work is done.

The next thing to be examined is the patient's pre-treatment occlusion. If this is all right, then, what the dentist will have to decide is whether the treatment can be done without changing the pre-existing occlusal condition of the patient. Accordingly, a conformative approach is adopted. If the occlusion has to be changed, then the dentist will need to take a re-organized approach and aim for an ideal occlusal condition. These two approaches make for good occlusal practice.

In order for a dentist to understand how the upper and lower jaw relate to each other, a bite registration needs to take place.

Factor #90: TMJ Exam

The temporomandibular joint or TMJ connects the jaw to the side of the head. In normal circumstances, this helps in talking, chewing and yawning. However, in cases of TMJ dysfunction (TMD), the patient experiences multiple symptoms.[163] These include:

- Stiffness in the jaw muscles
- Limited movement or locking of the jaw
- Problems in biting
- Painful clicking, popping or grating sounds in the jaw joint when opening and closing the mouth
- Pain in the area where the skull meets the jaw, which then spreads to the back of the head, and down the neck
- A change in the alignment of the upper and lower teeth
- Tension headache or migraine
- Vertigo, dizziness, and ringing in the ears

The pain is due to painful muscle spasms which are related directly to psychological stress. Initially, the pain can be handled by resting the jaw, massage of the chewing muscles, eating soft foods, avoiding chewy food and chewing gum, not opening the mouth too wide, using bite plates, applying heat, or taking simple analgesics. In case of injury, immediate treatment with ice is needed. Later heat therapy can be done. For tension headaches, tricyclic anti-depressants can be taken. In most cases, a night guard or orthotic appliance will treat the symptoms. In extreme cases, surgery might be needed. This disease tends to affect women more. Psychologists believe that this is because of the changing role of women in the world today. Most patients fall in the 20 to 40 year age group.

What happens is that the muscles which are used for chewing are affected by stress and become stiff. In some cases, to relieve tension, the patient might clench or grind the teeth. This aggravates the condition. Counseling helps in relieving stress and tension. Alternatively, the patient is taught relaxation techniques, including biofeedback. However, it would be best to go to the dentist as soon as possible and not suffer. Physical examination includes feeling the jaw joints and chewing muscles for any pain or tenderness; listening for popping or grating sounds during jaw movement; checking for limited motion or locking of the jaw while opening and closing the mouth. It is important to check the patient's medical and dental history to see if there has been a similar problem before. Dental x-rays can be done to evaluate the issue. However, to eliminate all doubt, a CT scan should be done.[164] It is extremely important to relieve the patient's suffering.

If you have TMJ disorder, make sure your dentist completes a thorough TMJ exam on you.

CHAPTER 9

FACTORS 91–100
TOP 10 COMPONENTS OF YOUR
DENTIST'S WEBSITE

Factor #91: Testimonial Videos

Choosing the dentist that a family or any person would like to go to is a difficult thing. The best dentists are tech-savvy and have their own websites. The more attractively informative the website, the more confident it can make you feel about your choice in dental care. One of the things that contribute to this is testimonial videos. Testimonials are endorsements of the service provided at the dental office. How a patient is made to feel from the time he enters the dental office, right up to the time he leaves it, which would include any dental procedures that he went through, ensure that this is the dental office he would choose in the future. This testimonial video can be put up on the dentist's website.

Testimonial videos about specific dental procedures, given in a manner that is easily understood, are effective. It is important for the person seeing the video to get an idea of the way the dentist goes about his work: whether he inspires confidence, what his attitude to the patient is, what equipment he uses, and how his assistants work. It is also possible to get a sense of the general ethos of the dental office. From the manner of talking, and the questions asked and answered, anyone seeing the video will know that they are viewing a live dental procedure, and that it has not been set up. Testimonial videos taken by satisfied patients helps build trust in potential patients.

The testimonial videos could be on one page, or specific services offered could be backed up by a testimonial video. Since a video testimonial is personal, it is both genuine and reliable and adds weight to the dentist's website.

Look for testimonials on your dentist's website.

Factor #92: Patient Forms

Patient forms are often available online from the website of the dentist the patient has chosen to go to. These forms are usually comprehensive and informative, and give the dentist an idea of the patient, the dental problem, and the medical history. The good thing about patient forms being available online is that you can fill in the form unhurriedly, in the privacy of your home.

When you go in for your appointment, you can bring a copy of the patient forms. You also bring all the documentation that the dentist might have asked for. The dentist verifies all the information in the forms and goes over all the details with you. If any corrections have to be made, they are made at this point, and the form is stored in the computer in a new file created for the patient. Digitizing the whole procedure makes it easily retrievable. Some dentists take the trouble to create an adjunct to the medical questionnaire.

Look for patient forms on your dentist's website.

Factor #93: Doctor's Biography & Qualifications

While looking around for the right dentist, you will visit the websites of many dentists. Web research has shown that the most visited page is the one where the dentist's biography and qualifications are given. Every patient wants to get the best dental care that they can afford, and they usually want the best dentist around. The dentist's biography would give information on where he grew up, which schools and colleges he went to, his family, his hobbies, his post-doctoral training, the awards he might have won in his field of specialization, whether he is progressive and committed to continuing education, and what professional memberships he has. In fact, just about everything that would help build an image of him in the mind of the patient. Often the page with the dentist's biography has videos and photographs of him on it.

Take a look at your dentist's biography and qualifications on his website.

Factor #94: Mission Statement

When choosing the right dentist for you, look at the dentist's Mission Statement on his website. That will help you understand what his philosophy is and what type of treatment style he has.

The dental office mission is the guiding light of the dental office and all who work there. The office mission outlines the fundamental purpose of the dental office, its philosophy and its values. It is only if there is a clear office vision that the dentist and his team will be able to work out their priorities. The office mission brings into focus the reason for existence of the dental office, and what specific needs of society it is going to fulfill. The office mission is used to maintain standards. It is also used like a compass to verify if the dentist and his team are on the right track. A powerful mission statement defines all the work that is being done, gives direction to the effort put in, ensures focus and commitment, and most importantly is deeply inspiring. A genuine office statement inspires confidence in the patients and they know that they are in safe hands. Pithy statements should be used for impact.

A mission statement is a live thing. Changes are constantly happening, and since the dentist and his team are committed to continued learning and keeping themselves abreast of all that is happening in the field of dentistry, they will need to modify the statement to reflect the changes.

A mission statement is not to be used as a marketing gimmick. It is, in fact, a reminder of the core competency of the members of the clinic. In case the dentist is putting together a team to start a new office, it would be good to involve the team members in the creation of the mission statement. The key features would definitely be to respect each patient and his specific dental needs, provide the best dental care possible using the latest technology, responsible diagnosis and assessment, treatment options, superior procedures, and excellent patient care. Other objectives could be: focus on lifetime family oral health care, punctuality, psychological comfort, painless experience, patient education, personalized care and personalized treatment, preventive dental care, building up of self-confidence and self-esteem, highest standards of quality dental care, pursuit of excellence through continuing education of all members of the team, complete satisfaction of the patient, responsibility for public oral health care, highest clinical standards and highest level of infection control, dental care in a relaxed and pleasant atmosphere, all dental concerns attended to by a team of highly trained, professional and ethical staff.

If your philosophy does not align with his, move on. He is not the right dentist for you. A mission statement gives you a sneak peak at the dentist's way of thinking, of his office's belief system, and of his purpose.

Take a good look at your dentist's mission statement on his website, and determine if that relates to your mission for your oral health.

Factor #95: Amenities and Technology Offered

Modern dentists concentrate on preventive oral health care. Most of today's dentists try to make their clinics attractive and friendly places. Everyone requires dental care and maintenance, and modern dental offices have the facilities for all kinds of dental treatment. The services offered are:

- A warm, welcoming office
- Hospitality, hot towels, eye and neck pillows, and magazines
- Education in oral health care including the right techniques for brushing and flossing
- Preventive care
- Thorough dental examination of each tooth
- Cleaning, scaling and polishing
- Dental x-rays
- Diagnosis and treatment of periodontal disease, and maintenance of the gums
- Laser periodontal therapy as well as antibiotic periodontal therapy
- Cosmetic dentistry which includes composite fillings, sealants, crowns, bridges, implants, veneers and tooth whitening/bleaching
- Tooth extractions
- Partial and full dentures including repairs and re-alignment of dentures
- Bone grafting
- Bite registration
- Root Canal therapy
- Composite Bonding
- Pediatric dentistry
- Geriatric dentistry
- Night guards
- Screening for oral cancer and the subsequent treatment
- Oral surgery

- TMJ treatment
- Cone Beam CT scanner and digital radiography
- Use of sedation while dental work is being done
- Use of headphones to listen to music while dental work is in progress
- Flat screen monitors
- Use of state-of-the-art technology for all dental procedures, such as intraoral and extraoral cameras, ultrasonic scaling units, Piezo Sonic techniques, Dental loupes, and Atraumatic forceps
- All dental records and images are stored in the computer
- Laser dentistry
- CAD/CAM technology
- Interactive patient education
- Some dental offices include a spa for a more complete experience. The dental spa offers dental services along with spa services such as aromatherapy, massage, eye masks and pillows, skincare, and relaxation techniques
- Cosmetic treatment for smile makeover
- Blankets and snuggies while undergoing treatment
- Sleep apnea appliances
- Regular comprehensive dental physicals
- Free initial consultations
- Counseling
- Online patient forms
- Informative and comprehensive website
- Well-planned offices, and provision for those in wheelchairs, or for those who are challenged in some way or the other
- Punctual appointments
- A knowledgeable, caring and courteous team
- Ability to give reliable referrals
- Advice about financial resources

You want to look for amenities and technology available to you from your dentist's website and find out if they have what you want.

Factor #96: Directions To The Office

The worst thing that can happen to you is to have a hard time finding the dental office. The dentist should give clear directions to the office on his website. Ensure that the correct address (preferably the mailing address) is mentioned on the website along with a Google Map or MapQuest link.

Certain points to remember are:

- Work out the simplest route, even if it takes a little more time.
- Give specific distances, such as the distance in kilometers and miles, how many blocks or streets they will pass, or how many traffic lights they will pass. Giving a *drop dead* landmark is a good idea because if the person who is driving arrives at this landmark, he will know he has passed the dental office.
- Give directions from all the major roads. In case the road has been re-named, use both the old name as well as the new name.
- In case there is one prominent landmark, use this as the point of reference and mark all the roads that lead from this point to the dental office.
- Note the mileage between points, if you are using certain landmarks as points of reference.
- Mark all turnings clearly and correctly i.e., in case you are using directions such as *turn left*, or *turn right*, then the direction of approach should also be mentioned, so that there is no confusion about left and right. Likewise for the cardinal directions (North, South, East and West).

- Use landmarks that people are familiar with. If the landmarks are linked, it makes it easier for the person driving to reach the destination.
- It is a good idea to check out the routes that you have mentioned by driving on them. Make any necessary changes.
- Give very precise and clear directions, in as few steps as possible.
- In case there are any confusing points on the road, mention them so that the person driving is prepared to deal with it.
- Hand-outs of a map, with landmarks and instructions, should also be kept in the dental office. Google Maps and MapQuest are good sites that offer detailed maps with directions.
- An important tip is that women use landmarks while men use direction and distance! *Women are from Mars and Men are from Venus*, right?

Factor #97: Videos about Procedures

In order to understand what your dentist may talk to you about, check out the procedure videos on his website. These videos show computerized versions of dental procedures, the reasons for them, how they are done, and how recovery occurs.

The various dental procedures that are done are shown in an animated form. Patients see how the dental treatment is done with the latest in technological advances, and hear the commentary that goes on explaining the whole procedure. This helps remove the fear of dental gadgetry from the mind of the patient. Sometimes, videos of the patients who underwent treatment in the dental clinic are shown with the condition of their teeth both before and after the treatment. This inspires confidence in the patients. Dentists feel that showing educational videos help to remind patients about the questions they wanted to ask about their own dental concerns. Dentists found it amazing how positively patients responded to their dental treatment, all because they had seen it on video. In fact, patients brought back their friends to the dental clinic because of their good experience. Since modern dentistry uses the technique of co-diagnosis, where the patient and the dentist work together to understand the problems and work out a treatment plan, using interactive educational videos is a great boon. The patient and dentist can pause the part they want to zoom in on, and then point and even draw on the screen.

If your dentist has procedure videos on his website, it really means he wants to give you the best care possible. It means he is transparent and doesn't want to hide anything from you, and that he values education.

Factor #98: Office & Doctor Contact Info

It is very important for patients to be able to contact the dental office and the doctor. If a dental issue or emergency arises, you want to be able to access the dental office or doctor as soon as possible.

Make sure your dentist's website has contact information for both the office and the doctor.

Factor #99: Educational Prevention Videos

Seeing a picture is more effective than words. This is the basis for educational videos. Modern dentists believe in preventive and comprehensive oral health care. As awareness of oral health care grows, patients want to know how to prevent dental issues from arising. While a visit to the dentist should be part of routine life, very often patients go to a dentist only when there is some problem. Nine times out of ten, the patient is quite scared of what might happen. It is precisely to remove such fears that dentists use educational videos—to educate patients in oral health care, as well as to show the patient how to prevent dental problems in the future.

Seeing and hearing real people immediately dispel all fears and apprehensions. These videos are short and attractively made. The whole structure of the mouth is shown in detail. The different kinds of teeth and the structure of the tooth are shown in a way that makes it easy to understand how important the teeth are. Dental infections and their cures are shown.

Factor #100: Professional Layout

A dental office's website ideally is a place where quality begins, and the atmosphere is relaxing and pleasant. The office's website should look like time and effort was put into it and should appear clean and organized.

A website is the face of a dental practice when people are looking for a dentist online. It should create a great first impression and it should mirror the impression one would get when walking into the dental office itself.

Look at your dentist's website. If it is sloppy and poorly constructed, more than likely the dentist's work is sloppy and poorly done as well. If it is beautiful and systematic, it is most probable that your dentist's work is beautiful and systematic as well.

Chapter 10

Factor 101
The MOST Important Factor in Choosing the Right Dentist

Factor #101: YOU!

YOU are the MOST important factor in choosing the right dentist. A visit to the dentist should be a pleasant event in today's day and age. You have the right to choose your own dentist. What do you look for when choosing the right dentist for you?

- Someone who you trust
- Someone who respects your confidentiality
- Someone who give you the right advice
- Someone who offers you dental diagnosis and painless treatment in a pleasant, relaxed and soothing atmosphere

You would want these basic requirements in your dentist because you have a healthy sense of self-respect and high self-esteem. Oral care is directly linked to physical well-being and so it is important that you do not neglect this.

Even if it is for a routine check-up, it is imperative to schedule regular visits to the dentist. This is besides the fact that in case there is an early problem, it can be detected at an early stage and treated successfully.

Regular and daily brushing with the right kind of toothbrush and fluoride-based toothpaste, and flossing ensure that you do not suffer from bad breath, and have healthy teeth and gums. Another habit to cultivate is to rinse the mouth after eating.

Good, well-cared for teeth ensure a great smile, and we all know how important a smile is for ourselves, as well as for those around us. It is good for your morale, will win you friends and spare you a great deal of anxiety.

Excellent oral care starts at home and continues in the dental office. Since you are the most important factor to maintaining oral health, it is imperative that you plan a wholesome and healthy diet. Snacking between meals is not good but if you feel you need it, eat nutritious snacks which are low in sugar.

Another very important reason to look after your teeth and gums is that often problems in the mouth are indicative of problems elsewhere in the body. Detected early, these can be treated, and you will continue to have a feeling of well-being.

If you look after your teeth when you are young, you will be sure to carry on with good oral health care as you grow older. This will greatly reduce dental problems that naturally occur because of age. It is a wonderful feeling to be complimented on having good teeth, no matter at what age.

Make sure you find the right dentist for you!

You deserve it!

Eat well and smile often!

Acknowledgements

In the Name of God, The Most Merciful, and The Most Kind. This book would not have been possible without the help of many people. Before thanking anyone, I have to thank God for giving me everything that I have, for my family, my faith, my profession, my education, and my health. I start this book in His name, and pray for peace, happiness, success, and education for everyone of you.

I have to thank YOU for purchasing and reading this book. You allow dentistry to exist and continue to improve. You have given me a forum to teach you what I love about dentistry and what I hope you deserve to receive from your dental experience. You have also told me how important your oral health is to you.

To my parents, Ahsan Haque and Seema Ahsan (my Abu and Ami): Thank you for raising me, for instilling the concept of hard work in me, for pushing me to be the best I can be, for educating me, for culminating my faith and morality, and for shaping me into the person that I am today.

To my loving wife, Siraj Haque: Thank you for putting up with me for the past decade, for supporting me, for your friendship, for your love, for your companionship, for giving me two beautiful and wonderful children, for joining me in my journey deeper into dentistry, for taking care of our family, and for everything that you do. Siraj, you do more than any man could ever ask for. Thank you! I love you.

To Maaz Haque and Sana Haque, my children: Thank you for giving me so much happiness and joy, for letting me teach you about the world and our faith, for our reading time, for sports, for inspiration, and for your beautiful smiles!

To my brother, Usman Haque: Thank you for helping me understand that anything is possible, for your hard work throughout your life, and for your love and support.

To my grandparents: Justice (retired) Zahoor Ul Haque, Shamima Khatoon, and Ghias Siddiqui. Thank you for giving me two amazing parents, and thank you for your constant love.

To the University of Connecticut School of Dental Medicine: Thank you for giving me the gift of a great dental education, for kicking me into gear, for the ongoing research that helps make dentistry better and better each day, and for giving me the opportunity to become a dentist. I was blessed to be taught by some of the finest dental professors, researchers, and educators in the world: Dr. Monty MacNeil, Dr. Edward Thibodeau, Dr. Peter Robinson, Dr. Thomas Taylor, Dr. Ravindra Nanda, Dr. David Shafer, Dr. Joseph Grasso, Dr. Frank Nichols, Dr. Clarence Trummel, Dr. Jeffrey Bennett, Dr. Jon Agar, Dr. Jacqueline Duncan, Dr. Martin Freilich, Dr. Kamram Safavi, Dr. Alan Lurie, Dr. Arthur Hand, Dr. Jon Meiers, Dr. Reza Kazemi, Dr. Michael Goupil, Dr. Howard Mark, Dr. George Paul, Dr. Ellen Eisenberg, Dr. Marion Frank, Dr. Ashraf Fouad, Dr. David Newitter, Dr. Sarita Arteaga, Dr. John Harrison, Dr. Joseph D'Ambrosio, Dr. Jon Goldberg, Dr. Robert Kelly, Dr. Dan Galindo, Dr. Jason Tanzer, Dr. Douglas Peterson, Dr. Larz Spangberg, Dr. Ra'ed Al-Sadhan, Dr. Richard Topazian, Dr. Elie Fernieni, Dr. Sunil Wadhwa, Dr. Flavio Uribe, Dr. Joe Zerella, Dr. Rajesh Lalla, Dr. Michael Brown, Dr. Ruth Goldblatt, and Mr. Joe Ficaro.

To Dr. Barry Polansky: Thank you for motivating me to write a book, for your vision and insight into dentistry, for your years of wisdom and talent, for your trail blazing, and for your service to all of your patients. Thank you for your mentorship, trust, and friendship.

To Dr. Arun Garg: Thank you for letting me know what is possible in dentistry, for training thousands of dentists and I in the field of implant dentistry, for your expertise, for your writings, and for your continued goal of making dentistry better. Most of all, thank you for helping me understand that anything is possible.

To Siraj Haque (now as office manager): Thank you for coming on board and transforming the office, for taking on the challenge, and for keeping me in check.

To my team at Oak Brook Smiles: LeeAnna Picchetti, Mariza Zapata, Brenna Schillinger, Cheryl Fontana, and Katie Tyree: Thank you for working so hard, for your sense of humor, for your warmth, and for contributing to the idea behind this book. You make our office amazing! Thank you for helping our office continue to grow, for helping me edit this book, and for your wonderful attitude. This book would not have been palatable without you! Thank you for opening your eyes to dentistry and for helping me put together the outline and game plan for this book. You will go far in dentistry!

To Dr. Bruce Howerton: Thank you for inspiring me to spend more time on my CT scans and write better reports. Thanks for showing me that ANYTHING is possible, and thanks for helping me with all of the CT factors.

To Dr. Bill Blatchford: Thank you for pushing me, for your advice on making my practice a better place to work in, and for helping me understand the potential that I have.

To Dr. Alfred "Duke" Heller, Dr. Arun Garg, Dr. Peter Dawson, Dr. John Cranham, Dr. Witt Wilkerson, Dr. Rhys Spoor, Dr. Paul Petrungaro, Dr. David Garber, Dr. Maurice Salama, Dr. Ronald Goldstein, Dr. Henry Salama, Dr. John Kois, Dr. Frank Spear, Dr. Joe Favia, and Dr. Chris Owens: Thanks for teaching me, for inspiring me, and for making dentistry my true passion.

To Dr. Irfan Atcha, Dr. Mohammed Shakeel, Dr. Vesna Sutter, Dr. Marcella Guzman, Dr. Anthony LaVacca, Dr. Stephen Spates, Dr. Steve Rhee, Dr. Darren Simpson, and Dr. Kaz Zymantas: Thanks for your passion for dentistry, your support, your guidance, and for pushing me to be better and better.

To Dr. James McAnally, Ed O'Keefe, Dan Kennedy, and Bill Glazer: Thank you for helping me and my practice think outside of the box, for your advice on making the experience in a dental office become fun.

To Dr. Arun Garg, Dr. Barry Polansky, Dr. Ravindra Nanda, Dr. Jeffrey Bennett, Dr. Vesna Sutter, Jeffery Gitomer, Anthony Robbins, Verne Hardish, and Nick Nanton: Thanks for giving me the inspiration to write a book.

To Bill King: Thanks for your motivation, inspiration, and dedication!

To Justin Jourdan, James Iwinski, Todd Colvin, Rudy Wolf, Chuck Vinyard, and Bill Brune: Thank you for your support, dedication, and brilliant ideas!

To Mark Val and his team at Premium Dental Lab, Jerry Ulaszek and his team at Artistic Dental Studio, and Luke Kahng and his team at LSK 121: Thank you for your spectacular lab work, dedication to dentistry, vision, commitment, and encouragement.

To Henry Schein and CAMLOG: Thank you for supporting me and helping me push myself to be the best I can be.

To all dentists: Thank you for all that you do for your patients day-in and day-out. Dentistry is a culmination of practice, science, medicine, and art. It is learned and improved upon every day, and is practiced by around 150,000 dentists in the U.S. Thank you for helping me try harder and harder every day.

References

1. Coppa, A., et al. Early Neolithic tradition of dentistry. *Nature.* 2006 Apr 6:755–756.

2. Matis BA, et al. Review of the effectiveness of various tooth whitening systems. *Oper Dent.* 2009 Mar-Apr;34(2):230-5.

3. Nanci, A. *Ten Cate's Oral Histology: Development, Structure and Function.* 2007. C. V. Mosby, St. Louis.

4. Listgarten MA, et al. Pathogenesis of Periodontitis. *Journal of Clinical Periodontology.* 1986;13,418-435.

5. Desvarieux M, et al. Periodontal Microbiota and Carotid Intima-Media Thickness: The Oral Infections and Vascular Disease Epidemiology Study (INVEST). *Circulation: The Journal of the American Heart Association.* 2005;111(5),576-82.

6. Ryan, ME, et al. The Influence of Diabetes on the Periodontal Tissues. *Journal of the American Dental Association.* 2003;134,34-40.

7. Shultis WA, et al. Effect of Periodontitis on Overt Nephropathy and End-Stage Renal Disease in Type 2 Diabetes. *Diabetes Care.* 2007;30(2),306-311.

8. Pitiphat W, et al. Maternal Periodontitis and Adverse Pregnancy Outcomes. *Community Dental Oral Epidemiolology*. 2008;36(1),3-11.

9. Polyzos, NP, et al. Effect of periodontal disease treatment during pregnancy on preterm birth incidence: a metaanalysis of randomized trials. *American Journal of Obstetrics and Gynecology*. 2009;200(3),225-232.

10. Mundy GR. Osteoporosis and Inflammation. *Nutrition Review*. 2007;65(12 Pt 2),147-151.

11. Posner M. Head and neck cancer. In: Goldman L, Ausiello D, eds. *Cecil Medicine*. 23rd ed. Philadelphia, Pa: Saunders Elsevier; 2007: chap 200.

12. Blot WJ, et al. Smoking and Drinking in Relation to Oral and Pharyngeal Cancer. *Cancer Research*. 1998;48, 3282-3287.

13. *National Comprehensive Cancer Network Clinical Practice Guidelines in Oncology: Head and Neck Cancers*. National Comprehensive Cancer Network; 2008. Version 2.2008.

14. Fedele S. Diagnostic Aids in the Screening of Oral Cancer. *Head Neck Oncology*. 2009 Jan 30; 1(1): 5.

15. Posner M. Head and neck cancer. In: Goldman L, Ausiello D, eds. *Cecil Medicine*. 23rd ed. Philadelphia, Pa: Saunders Elsevier; 2007: chap 200.

16. Rhodus NL. Oral Cancer and Precancer: Improving Outcomes. *Compend Contin Educ Dent*. 2009;30(8):486-498.

17. *National Comprehensive Cancer Network Clinical Practice Guidelines in Oncology: Head and Neck Cancers*. National Comprehensive Cancer Network; 2008. Version 2.2008.

18. Lockhart PB, et al. Poor Oral Hygiene as a Risk for Infective Endocarditis-Related Bacteremia. *Journal of the American Dental Association*. 2009;140(10):1238-1244.

19. Volchansky A, et al. Gingival Health in Relation to Clinical Crown Length. *Cases Journal*. 2009 Dec 23:2:9387.

20. Imrie DA. The Radiographic Appearance of Chronic General Periodontitis.*Proc R Soc Med*. 1928;22(1):74-75.

21. Erdemir EO. Evaluation of systemic markers related to anemia of chronic disease in the peripheral blood of smokers and non-smokers with chronic periodontitis. *European Journal of Dentistry*. 2008 Apr;2(2):102-109.

22. Beall AE. Can a new smile make you look more intelligent and successful. *Dent Clin North Am*. 2007 Apr;51(2):289-297.

23. Van der Geld P et al. Smile attractiveness. Self-perception and influence on personality. *Angle Orthod*. 2007 Sep;77(5):759-765.

24. Fitzgerald DB et al. Comparative cariogenicity of streptococcus mutans strains isolated from caries active and caries resistant adults. *J Dent Res*. 1977 Aug;56(8):894.

25. Ly KA et al. The potential of dental-protective chewing gum in oral health interventions. *J Am Dent Assoc*. 2008 May;139(5):553-563.

26. Quirynen M et al. Characteristics of 2000 patients who visited a halitosis clinic. *J Clin Periodontol*. 2009 Nov;36(11):970-975.

27. Douglass AB et al. Common dental emergencies. *Am Fam Physician*. 2003 Feb 1;67(3):511-6.

28. Haber J et al. The oral-systemic connection in primary care. *Nurse Pract*. 2009 Mar;34(3):43-8.

29. Pereira SM et al. Sugar consumption and dental health: Is there a correlation? *Gen Dent*. 2010 Jan-Feb;58(1):e6-e12.

30. Rafique S et al. Management of the petrified dental patient. *Dent Update*. 2008 Apr;35(3):196-204.

31. Ferrante L, Cameriere R. Statistical methods to assess the reliability of measurements in the procedures for forensic age estimation. *Int J Legal Med*. 2009 Jul;123(4):277-83.

32. Pruett HL. Listening to patients. *J Calif Dent Assoc*. 2007 Mar;35(3):182-5.

33. Vieira CL, Caramelli B. The history of dentistry and medicine relationship: could the mouth finally return to the body? *Oral Dis.* 2009 Nov;15(8):538-46.

34. Rafique S, et al. Management of the petrified dental patient. *Dent Update.* 2008 Apr;35(3):196-204.

35. Kravitz ND. The use of compound topical anesthetics: a review. *J Am Dent Assoc.* 2007 Oct;138(10):1333-9.

36. Palenik CJ. Environmental surface asepsis. *Dent Today.* 2005 Sep;24(9):122-124.

37. Rusmah M. Glutaraldehyde in dentistry--a review. *Singapore Dent J.* 1993 Jun;18(1):17-21.

38. Danhiez P et al. Use of the autoclave in dentistry. *Chir Dent Fr.* 1982 Jul 8;52(166):31-5.

39. Johnson LA, Schleyer TK. Developing high-quality educational software. *J Dent Educ.* 2003 Nov;67(11):1209-20.

40. White SC. Cone-beam imaging in dentistry. *Health Phys.* 2008 Nov;95(5):628-37.

41. Angelov N, et al. Periodontal treatment with a low-level diode laser: clinical findings. *Gen Dent.* 2009 Sep-Oct;57(5):510-3.

42. Kafas P, et al. Upper-lip laser frenectomy without infiltrated anaesthesia in a paediatric patient: a case report. *Cases J.* 2009 May 20;2:7138.

43. Scheutz F, Reinholdt J. Outcome of sterilization by steam autoclaves in Danish dental offices. *Scand J Dent Res.* 1988 Apr;96(2):167-70.

44. Chen C, et al. The effectiveness of an air cleaner in controlling droplet/aerosol particle dispersion emitted from a patient's mouth in the indoor environment of dental clinics. *J R Soc Interface.* 2009 Dec 23.

45. Covey, S. *The Seven Habits of Highly Effective People.* 2004:66-94. Free Press, New York.

46. Lundin, SC. *Fish! A Remarkable Way to Boost Morale and Improve Results*. 2000. Hyperion, New York.

47. Covey, S. *The Seven Habits of Highly Effective People*. 2004:204-235. Free Press, New York.

48. Blanchard KH, Johnson S. *One Minute Manager*. 1982:50-60. William Morrow and Co, New York.

49. Byrne R. *The Secret*. 2006. Atria Books, New York.

50. Ventegodt S, et al. Clinical holistic medicine: induction of spontaneous remission of cancer by recovery of the human character and the purpose of life. *ScientificWorldJournal*. 2004 May 26;4:362-77.

51. Puriene A, et al. Who is thought to be a "reliable dentist. *Stomatologija*. 2008;10(3):83-8.

52. Danforth RA, et al. 3-D volume imaging for dentistry: a new dimension. *J Calif Dent Assoc*. 2003 Nov;31(11):817-23.

53. Scarfe WC, Farman AG, Sukovic P. Clinical applications of cone-beam computed tomography in dental practice. *J Can Dent Assoc*. 2006 Feb;72(1):75-80.

54. Small BW. Cone beam computed tomography. *Gen Dent*. 2007 May-Jun;55(3):179-81.

55. Blot WJ, et al. Smoking and Drinking in Relation to Oral and Pharyngeal Cancer. *Cancer Research*. 1998;48, 3282-3287.

56. Fedele S. Diagnostic Aids in the Screening of Oral Cancer. *Head Neck Oncology*. 2009 Jan 30; 1(1): 5.

57. Coleton S. The use of lasers in periodontal therapy. *Gen Dent*. 2008 Nov-Dec;56(7):612-6.

58. Pesevska S, et al. Dentinal hypersensitivity following scaling and root planing: comparison of low-level laser and topical fluoride treatment. *Lasers Med Sci*. 2009 Jun 1.

59. Ludlow JB. Influence of CBCT exposure conditions on radiation dose. *Oral Surg Oral Med Oral Pathol Oral Radiol Endod.* 2008 Nov;106(5):627-8.

60. Apatzidou DA, Kinane DF. Nonsurgical mechanical treatment strategies for periodontal disease. *Dent Clin North Am.* 2010 Jan;54(1):1-12.

61. Lampe Bless K, et al. Cleaning ability and induced dentin loss of a magnetostrictive ultrasonic instrument at different power settings. *Clin Oral Investig.* 2010 Feb 3.

62. Hayashi M, et al. Influence of vision on the evaluation of marginal discrepancies in restorations. *Oper Dent.* 2005 Sep-Oct;30(5):598-601.

63. Van As G. Magnification and the alternatives for microdentistry. *Compend Contin Educ Dent.* 2001 Nov;22(11A):1008-12, 1014-6.

64. Danza M, Guidi R, Carinci F. Comparison between implants inserted into piezo split and unsplit alveolar crests. *J Oral Maxillofac Surg.* 2009 Nov;67(11):2460-5.

65. Vercellotti T. Piezoelectric surgery in implantology: a case report-- a new piezoelectric ridge expansion technique. *Int J Periodontics Restorative Dent.* 2000 Aug;20(4):358-65.

66. Yalcin S, et al. A technique for atraumatic extraction of teeth before immediate implant placement. *Implant Dent.* 2009 Dec;18(6):464-72.

67. Freedman G. Intraoral cameras: patient health through awareness. *Dent Today.* 2005 Jun;24(6):120, 122, 124.

68. Roshan S, Setien VJ, Nelson PW. Tooth whitening: a clinical review. *Tex Dent J.* 2008 Jul;125(7):602-13.

69. Boksman L. Current status of tooth whitening: literature review. *Dent Today.* 2006 Sep;25(9):74, 76-9; quiz 79.

70. Anxiety can increase post-surgical complications. *J Am Dent Assoc.* 1993 May;124(5):18.

71. Greenstein G. Therapeutic efficacy of cold therapy after intraoral surgical procedures: a literature review. *J Periodontol.* 2007 May;78(5):790-800.

72. Yucha C, Guthrie D. Renal homeostasis of calcium. *Nephrol Nurs J.* 2003 Dec;30(6):621-6.

73. Kritsidima M, et al. The effects of lavender scent on dental patient anxiety levels: a cluster randomised-controlled trial. *Community Dent Oral Epidemiol.* 2010 Feb;38(1):83-7.

74. McCaffrey R, et al. The effects of lavender and rosemary essential oils on test-taking anxiety among graduate nursing students. *Holist Nurs Pract.* 2009 Mar-Apr;23(2):88-93.

75. Arai YC, Ueda W. Warm steaming enhances the topical anesthetic effect of lidocaine. *Anesth Analg.* 2004 Apr;98(4):982-5.

76. Park SH, Mattson RH. Ornamental indoor plants in hospital rooms enhanced health outcomes of patients recovering from surgery. *J Altern Complement Med.* 2009 Sep;15(9):975-80.

77. Lohr VI, Pearson-Mims CH. Physical Discomfort May Be Reduced in the Presence of Interior Plants. *HortTechnology.* 2000 Jan; 10: 53-58.

78. Haviland-Jones J, et al. An Environmental Approach to Positive Emotion: Flowers. *Evolutionary Psychology.* 2005. 3: 104-132.

79. Schleyer TK, Johnson LA. Evaluation of educational software. *J Dent Educ.* 2003 Nov;67(11):1221-8.

80. Singer BA. Intraoral photography and videography: communicative roles in implant dentistry--Part I. *Dent Implantol Update.* 1996 Apr;7(4):28-31.

81. Nelson A, et al. The impact of music on hypermetabolism in critical illness. *Curr Opin Clin Nutr Metab Care.* 2008 Nov;11(6):790-4.

82. Wall M, Duffy A. The effects of music therapy for older people with dementia. *Br J Nurs.* 2010 Jan 28-Feb 10;19(2):108-13.

83. Harmat L, et al. Music improves sleep quality in students. *J Adv Nurs*. 2008 May;62(3):327-35.

84. Wang WC. Effects of tempo and other musical features on stress responses to college students. *J Acoust Soc Am*. 2010 Mar;127(3):1983.

85. Allred KD, et al. The effect of music on postoperative pain and anxiety. *Pain Manag Nurs*. 2010 Mar;11(1):15-25.

86. Ghosh S, Poddar MK. Higher environmental temperature-induced increase in body temperature: involvement of serotonin in GABA mediated interaction of opioidergic system. *Neurochem Res*. 1993 Dec;18(12):1287-92.

87. El-Gammal SY. Aromatherapy throughout history. *Hamdard Med*. 1990 Apr-Jun;33(2):41-61.

88. Braden R, et al. The use of the essential oil lavandin to reduce preoperative anxiety in surgical patients. *J Perianesth Nurs*. 2009 Dec;24(6):348-55.

89. Listgarten MA, et al. Pathogenesis of Periodontitis. *Journal of Clinical Periodontology*. 1986;13,418-435.

90. Imrie DA. The Radiographic Appearance of Chronic General Periodontitis.*Proc R Soc Med*. 1928;22(1):74-75.

91. Quirynen M et al. Characteristics of 2000 patients who visited a halitosis clinic. *J Clin Periodontol*. 2009 Nov;36(11):970-975.

92. Cranska JP. LANAP in the general practice: a case report. *Dent Today*. 2009 Dec;28(12):104, 106, 108.

93. Fontana CR, et al. Microbial reduction in periodontal pockets under exposition of a medium power diode laser: an experimental study in rats. *Lasers Surg Med*. 2004;35(4):263-8.

94. Angelov N, et al. Periodontal treatment with a low-level diode laser: clinical findings. *Gen Dent*. 2009 Sep-Oct;57(5):510-3.

95. Volchansky A, et al. Gingival Health in Relation to Clinical Crown Length. *Cases Journal*. 2009 Dec 23:2:9387.

96. Andreana S. The use of diode lasers in periodontal therapy: literature review and suggested technique. *Dent Today.* 2005 Nov;24(11):130, 132-5.

97. Cranska JP. LANAP in the general practice: a case report. *Dent Today.* 2009 Dec;28(12):104, 106, 108.

98. Borrajo JL, et al. Diode laser (980 nm) as adjunct to scaling and root planing. *Photomed Laser Surg.* 2004 Dec;22(6):509-12.

99. Paquette DW, et al. Locally delivered antimicrobials: clinical evidence and relevance. *J Dent Hyg.* 2008 Oct;82 Suppl 3:10-5.

100. Blot WJ, et al. Smoking and Drinking in Relation to Oral and Pharyngeal Cancer. *Cancer Research.* 1998;48, 3282-3287.

101. Fedele S. Diagnostic Aids in the Screening of Oral Cancer. *Head Neck Oncology.* 2009 Jan 30; 1(1): 5.

102. Land MF, Hopp CD. Survival rates of all-ceramic systems differ by clinical indication and fabrication method. *J Evid Based Dent Pract.* 2010 Mar; 10(1): 37-8.

103. Souza RA, et al. Importance of the diagnosis in the pulpotomy of immature permanent teeth. *Braz Dent J.* 2007;18(3):244-7.

104. Kawashima N, et al. Root canal medicaments. *Int Dent J.* 2009 Feb;59(1):5-11.

105. Koga DH, et al. Dental extractions and radiotherapy in head and neck oncology: review of the literature. *Oral Dis.* 2008 Jan;14(1):40-4.

106. Bortoluzzi MC, et al. Incidence of dry socket, alveolar infection, and postoperative pain following the extraction of erupted teeth. *J Contemp Dent Pract.* 2010 Jan 1;11(1):E033-40.

107. Kopp CD. Brånemark osseointegration. Prognosis and treatment rationale. *Dent Clin North Am.* 1989 Oct;33(4):701-31.

108. Avila G, et al. A novel decision-making process for tooth retention or extraction. *J Periodontol.* 2009 Mar;80(3):476-91.

109. Cordaro L, et al. Implant loading protocols for the partially edentulous posterior mandible. *Int J Oral Maxillofac Implants*. 2009;24 Suppl:158-68.

110. Maló P, Rangert B, Nobre M. All-on-4 immediate-function concept with Brånemark System implants for completely edentulous maxillae: a 1-year retrospective clinical study. *Clin Implant Dent Relat Res*. 2005;7 Suppl 1:S88-94.

111. Chiapasco M, et al. Bone augmentation procedures in implant dentistry. *Int J Oral Maxillofac Implants*. 2009;24 Suppl:237-59.

112. Lanigan DT, et al. Reconstruction of the atrophic mandible. *Ann Plast Surg*. 1986 Apr;16(4):333-53.

113. Jensen SS, Terheyden H. Bone augmentation procedures in localized defects in the alveolar ridge: clinical results with different bone grafts and bone-substitute materials. *Int J Oral Maxillofac Implants*. 2009;24 Suppl: 218-36.

114. Small BW. Cone beam computed tomography. *Gen Dent*. 2007 May-Jun;55(3):179-81.

115. Tetradis S, et al. Cone beam computed tomography in the diagnosis of dental disease. *J Calif Dent Assoc*. 2010 Jan;38(1):27-32.

116. Kleinman A, et al. Loma Linda guide: a stereolithographically designed surgical template: technique paper. *J Oral Implantol*. 2009;35(5):238-44.

117. Versteeg CH, et al. Efficacy of digital intra-oral radiography in clinical dentistry. *J Dent*. 1997 May-Jul;25(3-4):215-24.

118. Wenzel A. Digital radiography and caries diagnosis. *Dentomaxillofac Radiol*. 1998 Jan;27(1):3-11.

119. Geurtsen W. Substances released from dental resin composites and glass ionomer cements. *Eur J Oral Sci*. 1998 Apr;106(2 Pt 2):687-95.

120. Bergman M. Side-effects of amalgam and its alternatives: local, systemic and environmental. *Int Dent J*. 1990 Feb;40(1):4-10.

121. Puckett AD, et al. Direct composite restorative materials. *Dent Clin North Am.* 2007 Jul;51(3):659-75, vii.

122. Roshan S, Setien VJ, Nelson PW. Tooth whitening: a clinical review. *Tex Dent J.* 2008 Jul;125(7):602-13.

123. Meireles SS, et al. Efficacy and safety of 10% and 16% carbamide peroxide tooth-whitening gels: a randomized clinical trial. *Oper Dent.* 2008 Nov-Dec;33(6):606-12.

124. Lima DA, et al. In vitro assessment of the effectiveness of whitening dentifrices for the removal of extrinsic tooth stains. *Braz Oral Res.* 2008 Apr-Jun;22(2):106-11.

125. Boksman L. Current status of tooth whitening: literature review. *Dent Today.* 2006 Sep;25(9):74, 76-9; quiz 79.

126. LeSage BP. Minimally invasive dentistry: paradigm shifts in preparation design. *Pract Proced Aesthet Dent.* 2009 Mar-Apr;21(2):97-101.

127. Gürel G. Predictable, precise, and repeatable tooth preparation for porcelain laminate veneers. *Pract Proced Aesthet Dent.* 2003 Jan-Feb;15(1):17-24.

128. Brunton PA. Preparing anterior teeth for indirect restorations. *Dent Update.* 2004 Apr;31(3):131-6.

129. Felton DA. Edentulism and comorbid factors. *J Prosthodont.* 2009 Feb;18(2):88-96.

130. Marachlioglou CR, et al. Expectations and final evaluation of complete dentures by patients, dentist and dental technician. *J Oral Rehabil.* 2010 Feb 22.

131. Barotz C. How to make exquisite cosmetic dentures. *Dent Today.* 2002 Mar;21(3):54-6, 58-9.

132. Zini A, et al. The importance of correct diagnosis in complete dentures treatment. *Refuat Hapeh Vehashinayim.* 2009 Jul;26(3):15-9, 69.

133. do Amaral BA, et al. A clinical follow-up study of the periodontal conditions of RPD abutment and non-abutment teeth. *J Oral Rehabil.* 2010 Mar 10.

134. Ghavamnasiri M, et al. Porcelain fused to metal crown as an abutment of a partial denture: a clinical report. *J Contemp Dent Pract.* 2010 Mar 1;11(2):E064-70.

135. do Amaral BA, et al. A clinical follow-up study of the periodontal conditions of RPD abutment and non-abutment teeth. *J Oral Rehabil.* 2010 Mar 10.

136. Padmanabhan MY, et al. A comparative evaluation of agents producing analgo-sedation in pediatric dental patients. *J Clin Pediatr Dent.* 2010 Winter;34(2):183-8.

137. 1 Wood M. The safety and efficacy of intranasal midazolam sedation combined with inhalation sedation with nitrous oxide and oxygen in paediatric dental patients as an alternative to general anaesthesia. *SAAD Dig.* 2010 Jan;26:12-22.

138. Yagiela JA. Recent developments in local anesthesia and oral sedation. *Compend Contin Educ Dent.* 2004 Sep;25(9):697-706

139. Dionne RA, et al. Balancing efficacy and safety in the use of oral sedation in dental outpatients. *J Am Dent Assoc.* 2006 Apr;137(4):502-13.

140. Flanagan D, Goodchild JH. Comparison of triazolam and zaleplon for sedation of dental patients. *Dent Today.* 2005 Sep;24(9):64-6, 68-9.

141. Attanasio R. Bruxism and intraoral orthotics. *Tex Dent J.* 2000 Jul;117(7):82-7.

142. Cairns BE. Pathophysiology of TMD pain - basic mechanisms and their implications for pharmacotherapy. *J Oral Rehabil.* 2010 Mar 10.

143. Barghan S, et al. Cone beam computed tomography imaging in the evaluation of the temporomandibular joint. *J Calif Dent Assoc.* 2010 Jan;38(1):33-9.

144. Cooper BC, Kleinberg I. Relationship of temporomandibular disorders to muscle tension-type headaches and a neuromuscular orthosis approach to treatment. *Cranio*. 2009 Apr;27(2):101-8.

145. Bilyukov RG, et al. Obstructive sleep apnea syndrome and depressive symptoms. *Folia Med (Plovdiv)*. 2009 Jul-Sep;51(3):18-24.

146. Goodday R. Diagnosis, treatment planning, and surgical correction of obstructive sleep apnea. *J Oral Maxillofac Surg*. 2009 Oct;67(10):2183-96.

147. Dyken ME, Im KB. Obstructive sleep apnea and stroke. *Chest*. 2009 Dec;136(6):1668-77.

148. White DP. Central sleep apnea. *Med Clin North Am*. 1985 Nov;69(6):1205-19.

149. Weaver TE, Sawyer A. Management of obstructive sleep apnea by continuous positive airway pressure. *Oral Maxillofac Surg Clin North Am*. 2009 Nov;21(4):403-12.

150. Almeida FR, Lowe AA. Principles of oral appliance therapy for the management of snoring and sleep disordered breathing. *Oral Maxillofac Surg Clin North Am*. 2009 Nov;21(4):413-20

151. Haeger RS. How technology has transformed the one-visit initial exam. *J Clin Orthod*. 2004 Aug;38(8):425-34.

152. Lovegrove JM. Dental plaque revisited: bacteria associated with periodontal disease. *J N Z Soc Periodontol*. 2004;(87):7-21.

153. Savage A, et al. A systematic review of definitions of periodontitis and methods that have been used to identify this disease. *J Clin Periodontol*. 2009 Jun;36(6):458-67.

154. Haumschild MS, Haumschild RJ. The importance of oral health in long-term care. *J Am Med Dir Assoc*. 2009 Nov;10(9):667-71.

155. Al-Dakkak I. Diagnostic delay broadly associated with more advanced stage oral cancer. *Evid Based Dent*. 2010;11(1):24.

156. De Vos W, et al. Cone-beam computerized tomography (CBCT) imaging of the oral and maxillofacial region: a systematic review of the literature. Int *J Oral Maxillofac Surg*. 2009 Jun;38(6):609-25.

157. Neuman KA. Maximizing the use of an intraoral camera. *Dent Today*. 2003 Jul;22(7):72-5.

158. Terry DA, et al. Contemporary dental photography: selection and application. *Compend Contin Educ Dent*. 2008 Oct;29(8):432-6, 438, 440-2.

159. Pjetursson BE, Lang NP. Prosthetic treatment planning on the basis of scientific evidence. *J Oral Rehabil*. 2008 Jan;35 Suppl 1:72-9.

160. Zitzmann NU, et al. Patient assessment and diagnosis in implant treatment. *Aust Dent J*. 2008 Jun;53 Suppl 1:S3-10.

161. Squier RS. Jaw relation records for fixed prosthodontics. *Dent Clin North Am*. 2004 Apr;48(2):vii, 471-86.

162. Türp JC, et al. Dental occlusion: a critical reflection on past, present and future concepts. *J Oral Rehabil*. 2008 Jun;35(6):446-53.

163. Cairns BE. Pathophysiology of TMD pain - basic mechanisms and their implications for pharmacotherapy. *J Oral Rehabil*. 2010 Mar 10.

164. Barghan S, et al. Cone beam computed tomography imaging in the evaluation of the temporomandibular joint. *J Calif Dent Assoc*. 2010 Jan;38(1):33-9.

LaVergne, TN USA
31 January 2011
214558LV00001BA/3/P